T0273860

Lecture Notes, Worksheets, and Exercises for Basic Anatomy and Physiology

Martin Caon

Lecture Notes, Worksheets, and Exercises for Basic Anatomy and Physiology

Martin Caon
[Retired from Flinders University of South Australia]
Clarence Park, SA, Australia

ISBN 978-3-031-56295-2 ISBN 978-3-031-56296-9 (eBook)
https://doi.org/10.1007/978-3-031-56296-9

This Springer imprint is published by the registered company Springer Nature Switzerland AG
The registered company address is: Gewerbestrasse 11, 6330 Cham, Switzerland

Paper in this product is recyclable.

By the Same Author

Caon, M. (2020) *Examination Questions and Answers in Basic Anatomy and Physiology: 3rd ed. 2900 multiple choice questions and 64 essay topics*, Springer, 742p. ISBN: 978-3-030-47313-6, ISBN 978-3-030-47314-3 (eBook) https://doi.org/10.1007/978-3-030-47314-3

Caon, M. and Hickman, R. (2003) *Human Science: Matter and Energy in the Human Body 3rd ed.* Crawford Publishing House, Adelaide 472 p. ISBN 0 8633 3255 3

Abstract

A first-year university level course of introductory Human Anatomy and Physiology is presented as student's notes for 26 lectures that follow an organ-systems based approach. Some required basic chemistry and physics is also included. Many lectures have "additional material" which is provided for the student's interest but is excluded from the study course. Accompanying each lecture set are short-answer type revision questions that cover the important points of the lecture presentations. The answers may be constructed from the preceding lecture material. To ensure that the important points are not missed, extensive suggested answers to each of these 212 questions are provided and appear after the lecture material. Most of these questions have been used in end-of-course examinations. Also used in these examinations were the multiple-choice questions presented in a previous publication: Caon, M. (2020) *Examination Questions and Answers in Basic Anatomy and Physiology: 3rd ed., 2900 multiple choice questions and 64 essay topics*, Springer.

Contents

Lecture 1: Cells and Tissues

1 Cell Theory

The cell is the basic structural and functional unit of the body.
Procaryotes – bacteria (no nucleus or organelles except ribosomes)
Eucaryotes – plant and animal cells

2 Components of (Animal) Cells

1. **Plasma membrane**
2. **Cytoplasm** is the content of the cell, excluding the nucleus. **Cytosol** is the viscous gel-like fluid in which the organelles and inclusions are suspended and in which proteins, enzymes, ions and many other small molecules are dissolved. Many metabolic reactions occur in the cytosol.
3. **Organelles** – specialised structures of a characteristic shape that carry out specific roles in the cell.
   - **Mitochondria**: inner membrane folded into **cristae** (large surface area); they produce ATP (adenosine triphosphate); Krebs cycle occurs inside. They are able to replicate themselves.
   - **Lysosomes**: membrane-enclosed vesicles (formed in the Golgi) containing an acidic environment with enzymes capable of digesting (lysing) a wide variety of molecules.
   - **Centrosome**: contains two **centrioles** that function in a cell division.
   - **Ribosomes**: contain large amounts of RNA (ribonucleic acid); they synthesise proteins from amino acids.

M. Caon, *Lecture Notes, Worksheets, and Exercises for Basic Anatomy and Physiology*, https://doi.org/10.1007/978-3-031-56296-9_1

- **Nucleus**: largest organelle; a nuclear membrane with pores, which contains (one or more) **nucleolus** and **chromatin** (deoxyribonucleic acid (DNA) and associated proteins).
- (Multi-nucleate cells: muscle fibres, osteoclasts
- Cells without a nucleus: erythrocytes)
  - **Endoplasmic reticulum**: a system of membrane-enclosed channels continuous with the nuclear membrane (providing a large surface area); the **rough ER** is studded with ribosomes and stores newly synthesised molecules (by ribosomes); the **smooth ER** is the site of fatty acids, phospholipids and steroid synthesis;
  - **Golgi complex** (apparatus): four to six flattened sacs (cisterns) stacked on top of each other; it processes and delivers (via Golgi vesicles) lipids and proteins to the plasma membrane for secretion.
  - **Flagella and Cilia**: motile structures – sperm has a flagellum; the respiratory tract, the brain ventricles, some ducts in the testes and the fallopian tubes have ciliate cells.
  - **Cytoskeleton**: responsible for the movement of cells (e.g. phagocytes) and the movement of organelles and molecules within the cell and support and shape of the cell. **Microfilaments** – e.g. in muscle cells; **microtubules** – e.g. in the cilia.

3 Cell Membrane (Plasma Membrane)

It is a thin (6–10 nm) barrier separating the internal components of the cell from the exterior environment. It regulates the passage of substances into and out of the cell.

Fluid mosaic model of membrane structure: the membrane is a bilayer of lipids and has a mosaic of proteins "floating" (like icebergs) in a "sea" of lipids.

Lipids are **phospholipids** (75%), which form a *bilayer*, and **cholesterol** (20%). Proteins are **integral** (one end in the cell and the other end out of the cell) or **peripheral**.

Membrane proteins act as:

- Enzymes (catalyse reactions)
- Receptors (for signalling chemicals, e.g. hormones)
- Transporters (selectively allow entry to some solutes via channels (ATP-ases))
- Joiners (allow cells to adhere to each other)
- Recognisers (glycoproteins = identification tags)
- Attachment (to the cell's internal cytoskeleton and to extracellular structures)

4 Diffusion

All molecules and ions are in constant motion. In fluids, these particles move quite separately; the higher the temperature is, the faster is the motion. As particles approach each other, electrostatic forces cause them to repel. That is, particles are continually "bouncing off" each other as they move. Particles suffer millions of collisions per second. This continual *random* movement of particles amongst each other is called diffusion. It is the way particles move about inside and around cells (in the absence of specific transport mechanisms).

5 Transport of Ions and Molecules Through the Plasma Membrane

1. Diffusion of ions and molecules occurs along a concentration gradient (from Hi concentration to Lo concentration):
   (a) Lipid-soluble molecules (O_2, CO_2, NH_3, urea, alcohol, N_2, steroids, some vitamins) pass through the lipid bilayer.
   (b) Water-soluble molecules pass through pores formed by integral proteins (e.g. K^+, Ca^{2+}, Cl^-, HCO_3^-).
   (c) H_2O molecules pass rapidly through channels called aquaporins.
   Facilitated diffusion (of e.g. glucose) along a concentration gradient is assisted by specific protein carrier molecules in the membrane.
2. Osmosis: the net movement (diffusion) of **water** through pores (aquaporins) in a **selectively permeable membrane** along its concentration gradient.
3. Active transport: **requires energy** (from ATP) to move ions, amino acids and monosaccharides **against** their concentration gradient or charge gradient, e.g. the "Na^+-K^+ pump".

   (Secondary active transport: some substances, e.g. monosaccharides and amino acids, pass through the membrane using energy expended producing the Na^+/K^+ gradient.)
4. Bulk transport: a plasma membrane engulfs the substance, and the membrane-enclosed vesicle moves through the membrane. **Endocytosis** brings large substances (e.g. red blood cells (RBC), bacteria, proteins, polysaccharides) into the cell, and **exocytosis** takes large substances out of the cell *(e.g. at synapses, neurotransmitters are released from their vesicles by exocytosis).*

6 Tissues

Tissue: a group of similar cells (usually of common embryonic origin) that function together to carry out specialised activities
Biopsy: the removal of the sample of living tissue for staining and examination under the microscope
Histology: the microscopic study of tissues

Four Primary Types of Body Tissue

1. **Epithelial tissue**: (3% of the body) covers body surfaces; lines hollow organs, body cavities and ducts; and forms glands.
   - Cells closely packed (little extracellular material) and tightly bound together
   - One end of the cell is attached to the basement membrane (basal lamina) and the other end is the internal surface of the cavity or duct to a space
   - No blood supply (diffusion or absorption)
   - High rate of cell division
2. **Connective tissue (CT)**: (45% of the body) widely distributed tissue; basement membranes, bone, fat, blood, tendons, cartilage. It protects and supports the body and organs, binds organs together, stores energy as fat and provides immunity.
   - It consists of cells surrounded by an intercellular **matrix** (= ground substance and fibres).
   - The matrix may be fluid, semi-fluid, gelatinous, fibrous or calcified.
   - It has few cells and much matrix (extracellular material).
   - It is highly vascular (except cartilage)

   1. **Muscle tissue**: it consists of 50% of the body.
   2. **Neural tissue**: it forms 2% of the body.
3. **Epithelial Tissue (Epithelium)**
   (Layers of cells or glands)
   - **Simple squamous**: it consists of a <u>single layer</u> of flat (*squa*shed) cells, so diffusion through the layer takes place easily. It lines the heart, lymph and blood vessels (known as **endo**thelium). It is called **meso**thelium when in serous membranes.
   - **Simple cuboidal**: it forms a <u>single layer</u> of *cube*-shaped cells. The cells secrete or absorb, e.g. kidney tubules or the retina.
   - **Simple columnar**: it is a <u>single layer</u> of rectangular cells (*column* like). The cells secrete or absorb, e.g. line the GI tract, where some are goblet cells and some have microvilli.

- **Stratified transitional**: this consists of variable-shaped cells that are several layers thick and are able to stretch. The cells line the urinary bladder.
- **Stratified squamous**: it consists of several layers of cells for protection in areas of high wear. It is cuboidal to columnar in deep layers and squamous in superficial layers. Keratinised cells form the outer layer of the skin. Non-keratinised cells line the mouth, oesophagus, tongue, vagina and anus.
- **Pseudostratified columnar**: this forms a single "layer". All cells are attached to the basement membrane, but some do not reach the surface of the layer. The cells line the male urethra and upper respiratory tract. Some are ciliated, and some are goblet cells.
- **Glandular epithelium**: Exocrine glands secrete oil, wax, milk, sweat, tears, saliva and digestive enzymes into ducts. Endocrine glands secrete hormones, which diffuse into extracellular fluid and into the blood.

7 Connective Tissue

(*-blast* = an immature cell; it secretes the matrix *-cyte* = a mature cell.)

Nine (or So) Cell Types

- **Fibroblasts** (form fibres): they are always present in CT. They secrete hyaluronan and protein (= proteoglycans) and also secrete collagen, elastin and glycoprotein, which make up collagen fibres, elastic fibres and reticular fibres (respectively) in the matrix. Some become fibrocytes.
- **Chondroblasts**: in the cartilage, they secrete matrix and become chondrocytes.
- **Osteoblasts**: in the bone, they produce the organic matrix of bone and become osteocytes.
- **Haemocytoblasts**: in the bone marrow, they become red blood cells and white blood cells (WBC).
- **Adipocytes**: fat storage cells.
- **Leucocytes** (white blood cells): these function for defence. They include neutrophils and eosinophils (= microphages), basophils, monocytes and lymphocytes.
- **Macrophages** (develop from monocytes – a WBC): these are large cells that can engulf bacteria and cellular debris (perform phagocytosis).
- **Mast cells**: they produce histamine and heparin, which stimulate inflammation.
- **Plasmocytes** (develop from B lymphocytes – a WBC): they secrete antibodies and hence provide immunity.

Intercellular matrix composition – it has two components: **protein fibres** (collagen, elastin and reticulin) embedded in liquid and gel or solid **ground substance** (which contains a diversity of large molecules). The ground substance is amorphous and surrounds cells. It binds, lubricates, supports and provides a medium through which substances can diffuse.

8 Classification: Four Types of Connective Tissue

(In terms of the nature of the ground substance and the types and organisation of fibres)

1. Loose CT: packing material in space between organs
   (a) **Areolar**: widely distributed; contains many cells; mostly collagenous fibres; contains hyaluronic acid; around vessels, nerves and muscles; forms the synovial membrane, subserous fascia and superficial fascia (= hypodermis) layer under the skin
   (b) **Adipose**: fat cells (adipocytes) in loose CT; stores fat (triglyceride) and, hence energy; supports and protects organs
   (c) **Reticular**: with mainly reticular fibres; in the spleen, liver and lymph nodes

2. Dense CT: fibres occupy most of the volume.
   (a) **Dense regular**: shiny white appearance, tough and mainly collagen fibres arranged in a regular pattern; form tendons, ligaments, deep fascia and aponeuroses
   (b) **Dense irregular**: irregularly arranged collagen fibres; found in the dermis, heart valves, around cartilage (perichondrium), periosteum, fibrous capsules around kidney, liver, spleen, dura mater
   (c) **Elastic**: strong and stretchy; elastic fibres allow the tissue to stretch and snap back. Occur in lung, artery walls, vocal cords, vertebral ligaments, suspensory ligament of penis.

3. Supportive CT
   (a) **Cartilage**: very tough; it has no blood vessels or nerves. Cells are chondrocytes, which occur in spaces called lacunae. Perichondrium surrounds cartilage. e.g. Collagen and elastin fibres. It contains chondroitin sulphate in the ground substance.
      - *Hyaline cartilage* (gristle): the most common; bluish white; the ends of long bones, the nose, the trachea, and bronchi; flexible
      - *Fibrocartilage*: rigid; vertebral discs; menisci in the knee
      - *Elastic cartilage*: is elastic; found in ear lobe, epiglottis, eustachian tube.

   (b) **Bone**: it is rigid but not brittle. Osteocytes are in the lacunae and connected by canaliculi. Bone consists of: Inorganic calcium hydroxyapatite, collagen fibres and water matrix.

4. Liquid CT – has no fibres
   (a) **Blood**: erythrocytes (RBC), leucocytes (WBC) and platelets in a liquid matrix called plasma
   (b) **Lymph** = lymphocytes + interstitial fluid

9 Serous Membranes and Mucous Membranes

Serous membranes (serosa) consist of thin layers of areolar CT and are covered by mesothelium. They line internal body "cavities" and organs within the cavity. They secrete serous fluid, a lubricating fluid. Has parietal and visceral portions. e.g. **Pleura**, **peritoneum**, **pericardium**.

Mucous membranes (mucosa) consist of epithelium on top of CT (**lamina propria**) and are often surrounded by smooth muscle (**muscularis mucosae**). They line body cavities that open to the exterior (mouth, oesophagus, stomach, intestine). They secrete mucous to lubricate and prevent drying.

<u>Location of Systems Within Cavities</u>

The **dorsal cavity** contains the brain and spinal cord.
The **thoracic cavity** contains the heart and great blood vessels in the mediastinum, the lungs and some air passages.
The abdominal cavity contains the renal system and most of the digestive system.
The pelvic cavity contains the bladder, female reproductive organs and part of the large intestine.
Viscera = body organs of the abdomino-pelvic cavity
Abdomino-pelvic cavity = ventral cavity

<u>Structures Outside Body Cavities</u>

Skeletal system, integumentary system, skeletal muscle, kidneys (and adrenals), pancreas, duodenum, testes, sense organs, peripheral nerves and blood vessels

10 Additional Information

Mitochondria
In addition to energy production, mitochondria play a role in several other cellular activities. For example, mitochondria help regulate the self-destruction of cells (apoptosis). They are also necessary for the production of substances such as cholesterol and heme (a component of haemoglobin, the molecule that carries oxygen in the blood).

Mitochondrial DNA contains 37 genes, all of which are essential for normal mitochondrial function. Thirteen of these genes provide instructions for making enzymes involved in oxidative phosphorylation. The remaining genes provide instructions for making molecules called transfer RNAs (tRNAs) and ribosomal RNAs (rRNAs), which are chemical cousins of DNA. These types of RNA help assemble protein building blocks (amino acids) into functioning proteins.

The Human Body Described

The human organism is a complex arrangement of many different cells arranged into four types of tissues. These are epithelial, muscular, nervous and connective. Two or more tissue types that perform specific functions form a structure called an organ. A group of organs that act together to perform a particular body function is an organ system, of which there are 11. Together, all these organ systems maintain a living organism, such as a human being.

Organisational units (from smallest to largest): chemicals/molecules, organelles, cells, tissues, organs, organ systems, and human organisms.

Body Cavities

The body is made up of a number of "cavities" that are filled with organs and enclosed by a membrane. The body has two main cavities: the ventral cavity (which includes thoracic, abdominal and pelvic cavities) and the dorsal cavity (comprising the cerebral and spinal cavities).

Cavity Membranes

Organs within a cavity are surrounded by a serous membrane. The outer membrane is called the parietal membrane, and the layer in contact with the organs is known as the visceral membrane. The cerebrospinal membranes are called meninges.

Anatomical Directional Terms

It is necessary to be able to describe the location of any body structure in relation to another, in a language that everyone can understand (especially other health professionals). For this reason, a number of scientific DIRECTIONAL TERMS are used that refer to any area of the body.

For a body standing in the "**anatomical position**", the following pairs of opposing directional terms are used: anterior/posterior, superior/inferior, medial/lateral, proximal/distal and superficial/deep.

Abdominal Regions

The nine abdominal regions are named: left hypochondriac, epigastric, right hypochondriac, left lumbar, umbilical, right lumbar, left inguinal, hypogastric and right inguinal.

Body Planes

The body may be (theoretically) "sliced" in order to look at cross-sectional views and be able to recognise internal structures. Such views, produced by imaging machines, are used for diagnostic purposes. The terminology used to describe these body planes are frontal (or coronal), sagittal and transverse planes.

Regional Anatomical Terminology

It is useful to learn names of anatomical body regions as they form the stem words of many medical terms: cephalic, cranial, frontal, buccal, mental, facial, orbital, oral, acromial, axillary, brachial, carpal, palmar, mammary, umbilical, inguinal, pubic, phalangeal, lumbar, sacral, coxal, gluteal, femoral, popliteal, patellar, sural, crural, calcaneal, tarsal, pedal and plantar.

11 Cells and Tissues Homework Exercise 1

1. Name four of the organelles in a cell and describe their function.
2. Describe the structure of the plasma membrane (cell membrane).
3. Define the processes of "diffusion" and "osmosis".
4. What roles do proteins play in a cell's plasma membrane?
5. What is active transport?
6. Name and briefly describe the four types of tissue.
7. What are the functions of the epithelial tissue?
8. What is the difference between "loose" connective tissue and "dense" connective tissue?
9. From which of the four types of tissue are the following seven structures made:

   bone, lymph, tendon, cartilage, adipose tissue, glands and epidermis?

10. What structure separates the thoracic and abdominal cavities, and what is it made of?
11. What is the collective name for the contents of the ventral cavity?
12. What are the main functions of these membranes and the potential space they form?
13. What is the clinical condition that develops when air is able to enter the potential space of the pleural membrane?
14. What is the clinical condition called when the membrane of the abdominal cavity is inflamed?
15. How does an organ differ from a tissue?
16. Using the gastrointestinal tract as an example, list the cavity/cavities in which organs of this system are found.
17. Do all organs of the body lie within a body cavity? If not, give examples.
18. Using directional terms, describe the appearance of the body when it is standing in the "anatomical position".
19. Describe the position of each of the following using anatomical, directional terms: ear (compared to the nose and chin), elbow (compared to the wrist and shoulder) and vertebrae (compared to the sternum and kidneys).

Lecture 2: Review of Chemistry

(Text ref: Caon & Hickman 3rd ed pp 9–10, 53–54, 80–81 (sec 3.), 158–164, 170–180 and 224–2s26.)

1 Some Terms Defined

Most matter is usually "impure", being a mixture of variable amounts of several different substances.

Iron ore (is a mixture), iron (Fe) (an element), iron oxide (a compound of 2 elements), steel (an alloy), haemoglobin (a molecule with four atoms of Fe)

Element: refers to 90 naturally occurring simplest substances (listed in the "periodic table" – have *chemical symbols*)
http://www.chemicalelements.com/index.html
(or http://www.periodicvideos.com/)
Atom: the smallest particle of an element (contains protons, neutrons and electrons)
Proton: +vely charged subatomic particle in the nucleus of an atom (atoms of different elements have different numbers of protons)
Neutron: subatomic particle in the nucleus of an atom
Electron: −vely charged subatomic particle outside of the nucleus; tiny yet occupies the bulk of space in an atom
Chemical bond: occurs when the outer electron(s) of an atom participate with those of another atom in joining two or more atoms together to form a new substance (a "compound")
Metal elements (left-hand side (LHS) of the periodic table): always lose (donate) electrons in chemical reactions
Non-metal elements (right-hand side (RHS) of the periodic table): always gain electrons in chemical reactions
Compound: a substance formed when atoms from two or more elements are chemically combined in fixed proportions (have a formula, e.g. H_2O, $C_6H_{12}O_6$)
Molecular compounds: atoms forming the molecule are from non-metal elements, covalently bonded

M. Caon, *Lecture Notes, Worksheets, and Exercises for Basic Anatomy and Physiology*, https://doi.org/10.1007/978-3-031-56296-9_2

Covalent bond: the bond between two *non-metal* atoms in a molecule (the atoms share electrons, so BOTH *gain* electrons)

Molecule: the smallest particle of a molecular compound; consists of two or more atoms joined by covalent bonds

Ionic (non-molecular) compounds: formed when metal atoms (ionically) bond to six to eight surrounding non-metal atoms (and vice versa) – continuous crystal lattice structure.

Ionic bond: the attraction between a metal atom and **all** the surrounding non-metal atoms in the lattice (the non-metal atoms gain electron(s), while the metal atoms lose electron(s)

Ion: an atom that has gained (if a non-metal) or lost (if a metal) one (or more) electron(s) (when an ionic substance dissolves in water, ions separate and move about freely as **electrolytes**).

Molecules can be ions too!

Organic molecules: contain long chains of carbon atoms bonded to each other.

2 Suspensions and Solutions

These are mixtures in which one or more substances are dispersed throughout another.

Mechanical suspensions: the dispersed substance consists of relatively large particles that settle to the bottom quite quickly (e.g. medicines labelled "shake well before use", like Mylanta and injectable insulin).

Colloidal suspensions: particles are not large enough to settle quickly but are too large to pass through semi-permeable membranes (e.g. kaolin, proteins in the blood – produce **colloid osmotic pressure**)

Solutions: particles are so small that they never settle out and can often pass through S-PM

Solutions consist of:

(a) **Solvent** – usually liquid (water); the major component of the mixture
(b) **Solute(s)**: the minor component of the mixture; may be electrolytes, large ions or molecules of solid, liquid or gas (O_2, CO_2)

(**electrolyte** – a solid substance that will conduct electricity when dissolved in water. An electrolyte dissociates into ions, e.g. Na^+, Cl^-, K^+, Ca^{++}, NO_3^-, HCO_3^-, NH_4^+ PO_4^{--}, H_3O^+)

3 Solution Concentration

The concentration of a solution is a statement about the relative amounts of the solvent and solute present in the solution.

Per cent concentration: a common way of expressing solution concentration (e.g. on intravenous (IV) bags):

$$\%\text{Conc} = \frac{\text{Mass (g)}}{\text{Volume (ml)}} \times 100$$

Percentage concentration tells the number of grams of solute present in each 100 ml of solution.

For example, normal saline is 0.9% Na^+Cl^-, which is 0.9 g of sodium chloride per 100 ml of solution.

4 Density & Specific Gravity

$$\text{Density} = \frac{\text{Mass (g)}}{\text{Volume (ml)}}$$

Density has units of grams per millilitre.

As a solution becomes more concentrated, its density increases (because adding solute increases the mass much more than it increases the volume).

Specific gravity – density written *without* the units

(If urine SG = 1.009 ⇒ density = 1.009 g/ml.)

Urine specific gravity is a measure of urine's (an aqueous solution) concentration and may be measured with a clinical refractometer.

5 Working with Moles

The SI unit for the "amount of substance" (number of particles) is **mole** (symbol mol, 1/1000 of a mole = millimole = mmol)

To measure out the mass of a mole, you must:

(a) Know or find out the chemical formula of the substance; for example, glucose has the formula

$$C_6H_{12}O_6$$

(b) Know or find out (*from the periodic table*) the relative atomic mass (RAM) (= atomic weight) value for each type of atom present in the formula

$$\left(\text{Glucose}\,C_6H_{12}O_6\right)\text{RAMs are}\,C = 12, H = 1, O = 16.$$

(c) Multiply the RAM values by the number of atoms of each element appearing in the formula

$$12 \times 6 \ \text{atoms} = 72$$
$$1 \times 12 \ \text{atoms} = 12$$
$$16 \times 6 \ \text{atoms} = 96$$
$$\text{Total} = 180 = \text{relative formula mass}$$

A mole of any substance is a sample of the substance having a mass, in grams, equal to its relative formula mass.

Hence, a mole of glucose is an amount having a mass of 180 g.

For example, 1 mole of ammonia (NH_3) is:

$(1 \times \quad) + (3 \times \quad) = \quad$ g.

1 mole of water (H_2O) is:

$(2 \times \quad) + (1 \times \quad) = \quad$ g.

(RAM values: N = 14, O = 16, H = 1.)

A **mole** of any substance will always contain a fixed number of particles, and this is what makes the mole the most suitable unit for the measurement of the *amount* of any chemical.

1 mole is 6×10^{23} particles.

Molarity is solution concentration expressed in moles/L.

6 Osmosis

Osmosis is the diffusion of water molecules (H_2O) across a cell membrane from the side where the solution concentration is more dilute (i.e. where there are more water molecules) to the side where the solution concentration is greater (i.e. where there are fewer water molecules).

Osmosis always results in the more concentrated solution becoming more dilute.

This means that water molecules diffuse from where their concentration is high to where their concentration is low (NOTE: *concentration* of molecules ≠ number of molecules).

Water molecules cross the membrane by passing through pores (aquaporins).

If a solution is placed on one side of a semi-permeable membrane and pure solvent on the other, pure solvent will tend to move across the membrane into the solution. This movement can be prevented by applying hydrostatic pressure to the solution. The amount of pressure required to prevent a net movement of water molecules is the **osmotic pressure** of that solution.

Osmotic pressure (stated in mmHg or Pa) is a measure of the tendency of water to move into an aqueous solution. It is also another measure of solution concentration.

7 Tonicity

Isotonic solution: a solution with an osmolarity within the range of blood's osmolarity (280–300 mmol/l). When added to blood, an isotonic solution causes no net movement of water into or out of red blood cells.

Examples: 5% glucose, 0.9% sodium chloride and 9.5% sucrose (0.3% NaCl + 3.3% glucose)

Hypotonic solution: a solution with an osmolarity less than that of blood. It causes the movement of water into RB cells (causing them to swell and perhaps lyse).

Hypertonic solution: a solution with an osmolarity greater than that of blood. It causes the movement of water out of RB cells (causing them to shrink and crenate).

8 A Question!

What do 5% glucose, 9.5% sucrose and 0.9% sodium chloride solutions have in common that makes them all isotonic?

Let us compare 5% glucose with 9.5% sucrose.

$$5\% \text{ glucose} \rightarrow 5 \text{ g}/100 \text{ ml} \rightarrow 50 \text{ g}/\text{l}.$$

A mole of glucose is 180 g.
Therefore, 5% glucose contains
50 g ÷ 180 g = 0.28 mole of glucose in each litre of solution.

$$9.5\% \text{ sucrose} \rightarrow 9.5 \text{ g}/100 \text{ ml} \rightarrow 95 \text{ g}/\text{l}.$$

Sucrose has the formula $C_{12}H_{22}O_{11}$, and a mole of sucrose is 342 g (check this yourself!); therefore, 9.5% sucrose contains $\frac{95}{342} = 0.28$ mole of sucrose in each litre.

So, 5% glucose and 9.5% sucrose both contain the same number of moles per volume and therefore *the same number of particles per volume.*

Let us now look at sodium chloride.

$$0.9\%\ \text{NaCl} \rightarrow 0.9\ \text{g}/100\ \text{ml} \rightarrow 9\ \text{g}/\text{l}.$$

One mole of Na^+Cl^- is 58.5 g; therefore, 0.9% NaCl contains $9 \div 58.5 = 0.15$ mole of Na^+Cl^- in each litre of solution.

But remember! Na^+Cl^- is an ionic compound, and when it dissolves, it forms separate Na^+ ions and Cl^- ions!

Therefore, the number of moles of particles in each litre of solution is 0.30, which is close to that for 5% glucose and 9.5% sucrose.

So, isotonic solutions contain the same concentration of dissolved (non-penetrating) ***particles*****.**

To note, 5% glucose, 9.5% sucrose and 0.9% Na^+Cl^- solutions are all ***isotonic*****.**

9 The Osmole

Osmole: this is the amount of substance that must be dissolved in order to produce 6×10^{23} solute particles (be they ions or molecules) of whatever size.

For most covalent-molecular substances

No. of moles = no. of osmoles (e.g. glucose, sucrose).

For ionic substances

No. of osmoles = no. of moles × **no. of ions in the formula** (e.g. Na^+Cl^- has two ions
∴ osmolarity = 2 × molarity).

(Osmolality – the number of osmoles dissolved in each kilogram of water
Osmolarity – the number of osmoles dissolved in each litre of solution)

10 Definitions of Acid

Acid: substance that reacts with water producing H_3O^+ (hydronium) ions
Strong acid: reacts *completely* with water producing *large concentration* of H_3O^+, for example:

$$HCl + H_2O \rightarrow H_3O^+_{aq} + Cl^-_{aq}$$

Hydrochloric acid (100% ionised)
Weak acid: reacts *partly* with water producing a *relatively low* concentration of H_3O^+, for example:

$$CO_2 + H_2O \leftrightarrow H_2CO_3 + H_2O \leftrightarrow HCO^-_{3aq} + H3O^+_{aq}$$

Carbonic acid (only 0.2% ionised)
(However, H_2CO_3 is unstable, so it exists as CO_2. Carbonic anhydrase converts CO_2 to H_2CO_3 very quickly.)

11 Definition of Bases

Substances that react with water producing OH^-_{aq} (hydroxide) ions:

$$NaOH \rightarrow Na^+_{aq} + OH^-_{aq}$$

Strong base (sodium hydroxide)

$$NH_3 + H_2O \leftrightarrow NH^+_{4aq} + OH^-_{aq}$$

Weak base (ammonia)

12 Self-Ionisation of Water

$$H_2O + H_2O \leftrightarrow H_3O^+_{aq} + OH^-_{aq}$$

Pure water contains small, equal amounts of both $H_3O^+_{aq}$ and OH^-_{aq}.
∴ It is neutral (neither acidic nor basic).
When other substances are mixed with water, the concentration of $H_3O^+_{aq}$ or OH^-_{aq} may go up or down depending on the nature of the substance, but in all aqueous solutions:

$$\left[H_3O^+_{aq}\right] \times \left[OH^-_{aq}\right] = \text{constant} = 10^{-14} \ \text{mol / L}$$

For example, if $H_3O^+_{aq}$ conc. goes up, then OH^-_{aq} conc. must come down and vice versa.

That is, for pure water:

$$\left[H_3O^+{}_{aq}\right] = \left[OH^-{}_{aq}\right] = 10^{-7}\,mol/L$$

13 pH, Measure of Acidity ("Puissance" of Hydrogen)

pH = −log of $H_3O^+{}_{aq}$ conc. (concentration expressed in mol/L); hence, there is a factor of 10 difference in concentration between pH of 7 and 8.

$$\begin{aligned} pH < 7 &\quad \text{Acid} \\ pH = 7 &\quad \text{Neutral} \\ pH > 7 &\quad \text{Basic (alkaline)} \end{aligned}$$

pH of stomach juice is ~1.5.
pH of blood is 7.35–7.45.
At pH = 7.4, $[H_3O^+{}_{aq}] = 4.0 \times 10^{-8}$ mol/L
and $[OH^-{}_{aq}] = 2.5 \times 10^{-7}$ mol/L.

14 Buffer

Buffer is a solution that resists change to its pH. A buffer solution contains two components, a weak acid and its slightly basic salt. These components are "conjugate pairs". One component (the basic one) is able to remove hydronium ions from the solution; the other (the acidic one) is able to remove hydroxide ions from the solution.

1. H_2CO_3/HCO_3^- (bicarbonate buffer in blood)
2. $H_2PO_4^-/HPO_4^{2-}$ (phosphate buffer in cells)
3. Protein buffer (carboxylic acid group/amine group), e.g. plasma proteins in blood and haemoglobin in RBC.

15 Additional Information

Osmolality determines osmotic pressure, but osmolarity is more often stated as it is easier to express body fluid quantities in litres rather than kilograms of water (and for the dilute solutions in the normal human body, the difference is small).

Osmotic pressure (mm Hg) = 19.3 × osmolality milliosmole/kilogram of water.
The osmotic pressure of blood is about 5500 mm Hg = 732 kPa.
Of seawater = 27 atm!

Note:

- About 0.8 of the total osmolality of interstitial fluid is due to sodium and chloride ions.
- Around 0.5 of total intracellular osmolality is caused by potassium ions.
- The osmotic pressure of plasma in capillaries is ~20 mm Hg greater than that in interstitial fluid due to the effect of plasma proteins.
- "Calculated" mosmol/L is > "corrected" osmolar activity (mosml/L) because the ions and molecules exert an attraction (or repulsion) on each other that decreases (or increases) osmotic "activity".
- Stated osmotic pressures would be exerted if the fluids were on one side of a cell membrane and pure water on the other.
- Total osmotic pressure for plasma, 5450 mm Hg, is 19.3 times the corrected osmolality (282.5 mosmol/L).
- Any lack of osmolalic equilibrium within a cell is corrected within seconds to a minute (except if drinking is required, in which case it will take up to 30 mins for fluid to distribute via the gut and blood).
- *The osmolalities of extracellular and intracellular fluids remain exactly equal to each other except for a few minutes after a change in one of the fluids occurs.*
- *The number of osmoles of osmotically active substance in each compartment, in the extracellular fluid or in the intracellular fluid, remains constant unless one of the osmotically active substances moves through the cell membranes to the other compartment.*
- The tendency for osmosis to occur is expressed in pressure units because in some parts of the body, water also moves by filtration due to pressure differences. Hence, the net effect of water movement by osmosis and filtration can be assessed.

Hickman definition: isotonic solution = one that causes no net movement of water into or out of cells.

Isotonic – solutions with ~300 mosmol/L concentration of **non-penetrating** solutes
Iso-osmotic – solutions with ~300 mosmol/L concentration of solutes of **either a penetrating or a non-penetrating** nature
Non-penetrating solute – large polar organic molecules, plasma proteins and Na^+, Cl^-, and K^+ (these ions actually can pass through the plasma membrane but are actively pumped back and, hence, are "de facto" non-penetrating)
Penetrating solute – urea

Normal range in blood:

$[HCO_3^-]$ = 22–31 mmol/L.
$[H^+]$ = 35–42 nmol/L.

CO_2 produced in cells dissolves in water (blood) to produce carbonic acid.
Carbonic acid disassociates into bicarbonate ions and hydrogen ions.
Hydrogen ions are buffered by phosphate, plasma proteins and bicarbonate.

$$CO_2 + H_2O \Leftrightarrow H_2CO_3 \Leftrightarrow HCO_3^- + H^+$$

(Excess CO_2 is excreted by the lungs.)
(Excess H^+ is excreted by the kidneys.)

Calcium is used:

- In most stages of blood clotting
- To bind to troponin to remove tropomyosin, which exposes the actin-binding site in muscle contraction
- As intracellular messenger to signal synaptic vesicles to fuse with the pre-synaptic membrane
- For bone mineralisation

16 Lipoproteins in Human Plasma

	ρ	% lipid
Chylomicrons	<0.96	98–99
VLDL	0.96–1.006	90–93
LDL	1.006–1.019	79–89
HDL	1.019–1.21	43–67

Blood Lipids
- Total cholesterol
- Triglycerides
- High-density lipoprotein (HDL)-cholesterol
- Low-density lipoprotein (LDL)-cholesterol

HDL sub-fractions (HDL2 and HDL3)
LDL sub-fractions (total apo-B, LDL apo-B, LDL-c/LDL, apo-B ratio, LDLRf)

What Is a Logarithm?
Numbers can be written as a power of 10.

For example, in $20 = 10^{1.3010}$, 1.3010 is the “power of 10” that produces the number 20.

Log_{10} of a number is that power of 10 that would produce that number. For example:

$1 = 10^0$, so $\log(1) = 0$.
$10 = 10^1$, so $\log(10) = 1$.
$100 = 10^2$, so $\log(100) = 2$.
$1000 = 10^3$, so $\log(1000) = 3$.
$20 = 10^{1.3010}$, so $\log(20) = 1.3010$.

$4 = 10^{0.6026}$, so $\log(4) = 0.6026$.

$\log(10^{-8}) = -8$.
$0.00000004 = 4.0 \times 10^{-8} = 10^{-7.4}$.

$\log(4.0 \times 10^{-8}) = \log(4.0) + \log(10^{-8}) = 0.6026 - 8 = -7.4$.

Calculator Example
Enter:
10, x^y,
8, +/−,
×, 4, =, (4×10^{-8})
Log, =, (−7.4)
So, $\log(4 \times 10^{-8}) = -7.4$.

17 Basic Chemistry Revision: Homework Exercise 2

1. Write out the definition of each of the following 15 things:

Element:
Atom:
Proton:
Neutron:
Electron:
Chemical bond:
Metal element:
Non-metal element:
Compound:
Molecular compounds:
Covalent bond:
Molecule:
Ionic (non-molecular) compounds:
Ionic bond:
Ion:

2. Identify the name of the elements and the number of atoms in the following:

(a) $C_6H_{12}O_6$
(b) CH_3COOH
(c) $NH_4^+ OH^-$

3. Which of the following compounds are covalent, and which are ionic?

(a) $C_6H_{12}O_6$
(b) CH_3COOH
(c) $Na+ Cl^-$
(d) $K+ Cl^-$
(e) $Ba++ SO_4^{--}$

Chemistry Calculations:

1. By following the steps below, calculate the mass of substance per 100 ml for an IV solution that contains 0.3% Na^+Cl^- and 3.3% glucose.

   If a solution has a concentration of *x*%, then there is *x* gram of substance per 100 ml of solution.

(a) What mass of sodium chloride is in 100 ml of 0.3% Na^+Cl^-?

____

(b) What mass of glucose is in 100 ml of 3.3% glucose?

____

(c) Now add the two masses together to get the total mass dissolved in 100 ml. of a 0.3% Na^+Cl^- and 3.3% glucose solution.

2. Perform a similar calculation to find the mass of substance per 100 ml for a 0.224% $K^+ Cl^-$ solution.
3. Perform the following steps to calculate the number of millimoles per litre in a 5% glucose solution.

   (a) First calculate the mass of glucose in 1 L:

   $$\text{mass}(g)\text{in}100\text{ml} = -\_\_\_\_\_\_\_\_ g$$

   (b) Multiply this answer by 10 to get grams per litre:

   $$\_\_\_\_\_\_\_\_\times 10 = \_\_\_\_\_\_\_\_ \text{g per litre}$$

   (c) Calculate the number of moles in this many grams of glucose in two steps:

   (i) Number of grams in one mole of glucose $C_6H_{12}O_6$ (RAMs are C = 12, H = 1, O = 16). Multiply the RAM values by the number of atoms of each element appearing in the formula:

   $$(12\times\_\_\_)+(1\times\_\_\_)+(16\times\_\_\_)=\_\_\_+\_\_\_+\_\_\_=\_\_\_\_\_\_$$

   The answer is the "relative formula mass".

   (ii) Divide the number of grams per litre [from b) above] by the relative formula mass [from c) part (i)]:

   $$\_\_\_\_\_\_\_\_\_\_\_\_ \text{mole}/1$$

   The answer is the number of moles per litre (it should be a decimal number less than 1.0).

   (d) Multiply the answer in (c) part (ii) by 1000 (shift the decimal three places to the right) to get the answer in mmol/l:

   $$\_\_\_\_\_\_ \text{mol}/11000 \_\_\_\_\_\_ \text{mmol}/1$$

4. Calculate the number of millimoles per litre in 1 L of IV bag of normal saline (0.9% $Na^+ Cl^-$).
5. Give the definition of an osmole.

   (a) What are the particles in the glucose solution called?
   (b) What are the particles in the saline solution called?

6. Use the number of moles per litre calculated in Q3 to state the number of osmoles per litre in 1 L of IV bag of 5% glucose (a covalent molecular substance ← hint!).

_________ moles / litre of glucose = _________ osmoles / litre of glucose

7. Calculate the number of osmol/l in a 1 L bag of 0.9% Na+ Cl- (normal saline – an ionic substance).
8. Characterise the following solutions as hypotonic, isotonic or hypertonic to blood plasma:

   (a) 0.9% sodium chloride
   (b) 0.3% sodium chloride
   (c) 5% glucose
   (d) 4% glucose
   (e) a solution containing 3.3% glucose and 0.3% sodium chloride
   (f) a solution containing 4% glucose and 0.18% sodium chloride
   (g) 9% sodium chloride

9. (a) What is the pH of a solution that contains H+ at a concentration of 10^{-5} mol/L?
   (b) Is the solution in (a) acidic or basic?
   (c) What is the pH of a solution that contains H^+ at a concentration of 3.2×10^{-5} mol/L?

Lecture 3: The Integument

The **skin**, **hair**, **nails**, **glands** and several **specialised receptors** make up the integumentary system.

Functions of the Skin

1. Body temperature regulation
2. Protection
3. Perception of stimuli
4. Excretion
5. Synthesis of vitamin D (calcitriol)
6. Immunity
7. (Social function: makeup, piercing, tattoos, keloid scars)

1 Structure of the Skin

About 7% of body weight, from 0.5 (eyelids) to 4.0 (heels) mm thick

Epidermis (outer layer): composed of epithelial cells and a dead layer

Dermis (inner layer): composed of fibrous connective tissue and exocrine glands

(*Hypodermis (or "superficial fascia"): composed of adipose (and some areolar connective) tissue;* ***not*** *part of the skin; stores fat, anchors skin to the muscle and allows the skin to slide over the muscle*)

1. **Epidermal Cells (Four Distinct Types)**

   (a) Basal cells (stem cells that divide to form keratinocytes). Keratinocytes are tightly connected to each other (by desmosomes), produce **keratin** (a fibrous protein).

M. Caon, *Lecture Notes, Worksheets, and Exercises for Basic Anatomy and Physiology*, https://doi.org/10.1007/978-3-031-56296-9_3

(b) Melanocytes have numerous branching processes that touch all keratinocytes in the basal layer; they produce **melanin** (*a pigment that protects against ultraviolet (UV)*), which is transferred to nearby keratinocytes.
(c) Dendrocytes (Langerhans cells and Granstein cells) arise from the bone marrow and migrate to the epidermis. They interact with lymphocytes (T cells) to assist in the immune response against micro-organisms and skin cancer. They are macrophages.
(d) Merkel cells are associated with a sensory nerve ending (called Merkel disc) for fine touch and pressure.

The epidermis is NOT VASCULARISED! Nutrients reach epidermal cells by **diffusion**.

2 Strata (Layers) of the Epidermis

1. Stratum germinativum/basale: it is the deepest layer, attached to the underlying dermis by protruding epidermal ridges (epidermal ridges push the overlying epidermis into "fingerprints"). It is single-cell thick and the youngest rapidly dividing keratinocytes. New cells push up the older ones above. Of the layer, 10–25% are melanocytes.
2. Stratum spinosum: this layer is 8–10 cells thick. The cells contain keratin filaments that span cells, attach to desmosomes and so hold the cells together. *Blisters (from repeated rubbing) are fluid-filled pockets in the stratum spinosum caused by the disruption of junctions between the cells – the layers of the skin separate.*
3. Stratum granulosum: it is a layer 3–5 cells thick. It contains granules that aggregate keratin and contains waterproofing **glycolipid***. Cells flatten and dehydrate, and nuclei organelles disintegrate (**glycolipids are lipids that contain carbohydrates, usually simple sugars like glucose*).
4. (Stratum lucidum (*only present in thick skin – palms, soles*): it consists of a few rows of clear, dead and flattened keratinocytes.)
5. Stratum corneum: it is the most superficial layer, 20–30 cells thick – up to ¾ of the thickness of the epidermis. It consists of dead, flat cells filled with keratin and is durable and expendable. *It protects the body against heat, light, chemicals, bacteria and abrasion. Cholesterol and glycolipid between the cells waterproof this layer. Burn victims lose much water because the skin is absent.*

2. **<u>Dermis</u>**

The dermis contains hair follicles, oil and sweat glands, nerve fibres, blood and lymph vessels.

It is a strong flexible connective tissue containing collagen, elastin and reticular fibres. (The dermis of animals can be tanned to make leather.)

Papillary Dermal Layer

This layer has folds and ridges called papillae that push up into the epidermal layer. It contains capillaries and lymphatics that supply the skin and sensory neurons.
The papillae contain capillary loops and light touch receptors – Meissner's corpuscles.
Dermal ridges – produce our fingerprints

Reticular Dermal Layer (is deep to papillary layer)

It is 80% of the thickness of the dermis. It contains Pacinian corpuscles sensitive to deep pressure.
It consists of interlacing collagen fibres arranged in parallel bundles, giving the skin strength and resilience (lines of cleavage).
Elastin fibres provide the stretch/recoil properties of the skin.
Flexure lines (occur at bone joints).

3 Glands of the Dermis

Sebaceous (oil) glands are found all over the body (except the palms and soles). They secrete **sebum** into hair follicles and keep the skin and hair from drying out. They contain triacyl glycerides, cholesterol, proteins, electrolytes and cell fragments.
Sebum inhibits the growth of bacteria.
Blocked ducts – produce "whiteheads", and "blackheads"

4 Sudoriferous (Sweat) Glands

Eccrine (merocrine) sweat glands are coiled tubes in the dermis discharging through the duct to the pore on the skin.

Sweat is a hypotonic filtrate of blood. It contains 99% water, Na^+Cl^-, vitamin C, dermicidin (a peptide with antibiotic properties), urea, uric acid and K^+ (pH = 4–6.8).
These glands have a function in thermoregulation under the control of the sympathetic division of the autonomic nervous system (NS).

Apocrine sweat glands are found in the axillary, nipple and anogenital areas. They function after puberty.

Ducts empty sweat + fatty substances and proteins into hair follicles.

Ceruminous glands in the ear canal secrete **cerumen** (ear wax).
Mammary glands (under the influence of pituitary hormones) secrete **milk**.
About 100 sweat glands, 15 oil glands, 230 sensory receptors per cm^2 of skin.

Hair consists of keratinised cells, and it has a protective function. The arrector pili muscle stands the hair up when cold (goosebumps).

Nails are modified epidermis consisting of keratin, useful for scratching and picking things up. A rich bed of capillaries underlies the nails (in the dermis).

5 Burns

First degree: only the epidermis is damaged (sunburn); it heals in 2–3 days.
Second degree: it injures the epidermis and upper dermis; blisters appear (partial-thickness burns). The skin regenerates (1–2 weeks) with little or no scarring.
Third degree (full thickness): it involves whole thickness of the skin. The nerve endings are destroyed. Extensive burns cannot heal themselves. Skin grafting is required to avoid fluid loss and infection.

If >10% of skin the area is full-thickness burnt ➔ critical.
If >20% of the skin area is burnt ➔ life-threatening.

6 Ageing of the Skin

Newborn skin is thin. During infancy and childhood, skin thickens and subcutaneous fat is deposited. During adolescence, the skin and hair become oilier; acne may appear. After 30 years, the skin begins to show the effects of "environmental assault". With old age, the rate of epidermal cell replacement slows, the skin thins and glands secrete less. Elastin fibres clump and degenerate, collagen fibres become fewer and stiffer, and the hypodermal fat layer diminishes → wrinkling. The number of active hair follicles diminishes → hair thinning.

7 Drug Administration

Topical: patted or wiped onto the skin, absorbed through the skin and *acts locally* (e.g. corticosteroids)
Transdermal: absorbed through the skin, *acts systemically* (e.g. nicotine patch, hormone replacement therapy (HRT), nitroglycerin, scopolamine patch, transdermal oxycodone patch)
Intradermal: injections of <1 cm^3 into the dermis
Subcutaneous: injected into the subcutaneous fat (hypodermis) layer

Lipid-soluble substances (e.g. steroids) can penetrate the skin because glycolipids between the cells of the stratum corneum present no barrier.

8 Sensory Receptors in the Skin

Tactile Receptors

1. Merkel cells monitored by sensory terminals called tactile discs: in the basal layer of hairless skin, detect fine touch
2. Meissner's corpuscles: for fine touch in the dermal papillae
3. Pacinian (lamellated) corpuscles: for deep pressure in the reticular layer
4. Ruffini corpuscles: for pressure and distortion in the skin (in the reticular layer)
5. Root hair plexus: to detect the movement/distortion of hair follicles
6. Free nerve endings between the epidermal cells: have small receptive fields for touch sensation

Thermoreceptors

1. Free nerve endings for heat: detect heat
2. Free nerve endings for cold: detect cold

Nociceptors

Pain receptors are free nerve endings with large receptive fields.
(Three types: extremes of temp, mechanical damage, dissolved chemicals – strong stimuli excite all 3!
➔ "burning sensation" is excited by many stimuli)

Vitamin D
= Calcitriol – required for the uptake of Ca^{++} from the gut

In the skin: UV radiation causes "modified cholesterol molecules" (cholecalciferol = provitamin D3) to be converted to "vitamin D precursor".
In the liver: enzymes modify this precursor (by adding a hydroxyl group [-OH]).
In the kidneys: "modified precursor" is used to produce (add a second hydroxyl group) and release calcitriol (vitamin D3) upon being signalled by the parathyroid hormone (PTH). So "vitamin" D3 is not necessarily a part of the diet but is actually a "hormone", so the skin can be considered an endocrine organ!

9 Additional Information

Human beta-defensin 2 is a natural anti-microbial peptide produced by keratinocytes when injured. It is part of the skin's natural defence mechanism (against atopic dermatitis and psoriasis).

Malignant melanoma incidence is ~20 cases/100,000 individuals. The majority are cured by surgery.

Patients with stage IV disease have a dismal prognosis. Metastatic melanoma patients' median survival is between 6.4 and 9.1 months. (Dacarbazine is the only FDA-approved drug!)

Non-malignant Skin Cancers+

1. Superficial basal cell carcinoma
2. Nodular basal cell carcinoma – surgically removed with scarring
3. Squamous cell carcinoma

Actinic keratoses (solar keratoses = sun spots) are reported in 40% of the Australian population! They can progress to squamous cell carcinoma. Treatment is cryotherapy with liquid nitrogen or CO_2 slush, surgical removal or topical therapy with fluorouracil cream.

10 Integument Revision: Homework Exercise 3

1. List the components of the integument.
2. What are the principal functions of the skin?
3. Describe the five layers of the epidermis.
4. List the types of cells in the epidermis. How do they differ?
5. Describe the structure of the two strata in the dermis.
6. What are the two types of glands in the skin, and what is their purpose?

Lecture 4: Thermoregulation and Homeostasis

(Caon & Hickman 3rd ed sect 5.10 and 5.11)

Physical Means of Heat Gain/Loss:

Radiation, convection, conduction, evaporation of water (perhaps ingesting hot liquids?)

Biological Means of Heat Loss/Gain:

Basal metabolism (BMR); sweating; shivering; skeletal muscle activity; subcutaneous fat insulation; vasoconstriction or vasodilation; eating, which increases liver activity; eliminating; and urinating.

1 Behavioural Temperature Control

- Seek an environment of appropriate temperature.
- Adapt the level of physical activity.
- Ingest hot/cold liquids.
- Adjust the level of clothing/exposed skin.

(Skin temp may be 32 °C when room temp is 24 °C.) Body core is maintained at ~37 °C (36.5–37.5), 0.4 °C less if measured orally. Why?

If not, chemical reaction rates and blood viscosity change, denaturation of protein occurs (convulsions at 41 °C)

Hyperthermia >38 °C (may be due to exercise)
Hypothermia <35 °C

When the capillary beds of the dermis are flushed with blood (at 37 °C), heat is transferred from the core to the epidermis and can be lost from the body (by radiation, conduction, convection, evaporation).

M. Caon, *Lecture Notes, Worksheets, and Exercises for Basic Anatomy and Physiology*, https://doi.org/10.1007/978-3-031-56296-9_4

Heat transfer to the skin is controlled by the degree of vasoconstriction of the arterioles and arterio-venous anastomoses (= capillary network) that supply blood to the venous plexus of the skin. Vasoconstriction is controlled by the sympathetic nervous system.

Vasoconstriction → skin temp approaches the temp of the surroundings.

Heat energy is lost from the body by:

(a) Evaporation of sweat, breathing
(b) Radiation, convection and conduction
(c) Eliminating, urinating

2 Conduction

When two objects touch, the one at the higher temperature will pass energy to the one at the lower temperature. **The energy transferred is known as heat**.

The particles of the object at the higher temperature are, on average, vibrating faster than the particles of the object at the lower temperature.

Each particle passes energy to the particle next to it *by contact*. The faster-moving particles of the high-temperature object collide with the slower-moving particles of the low-temperature object (and vice versa!). The result will be an increase in speed (and kinetic energy) for the slower-moving particles and a decrease in speed (and kinetic energy) for the faster-moving particles.

When heat is transmitted through a solid body in this way, the process is called **conduction**. A "warm" human body will heat "cold" hands (i.e. hands at a lower temperature than the skin of the body) placed in contact with it by conduction; a hot-water bottle in bed also transfers heat by conduction.

Clinical thermometers gain heat by conduction when applied to a patient.

Substances differ quite markedly in the ease with which they allow conduction to occur.

The skin, subcutaneous tissue and fat are heat insulators for the body. They maintain the core temp while allowing the skin temp to approach the temp. of the surroundings.

(At an ambient temp of 24–29 °C, most resting unclothed humans are in thermal equilibrium – metabolic heat produced is lost to the environment.)

Adipose tissue of the hypodermis insulates the body against heat loss as fat is 1/3 as conductive as muscle.

Each millimetre of fat insulates to the extent that a person can feel comfortable in a 1.5 °C lower temperature.

Most good conductors are metals or metal alloys (steel k = 400, aluminium k = 2100, copper k = 3900, silver k = 4200). Because these substances easily conduct heat, they "feel cold" when in contact with our skin. They can withdraw heat rapidly from us, and its loss produces the sensation of "coldness" even though the metal is at room temperature. For this reason, the nurse should heat the bedpans and other metal objects (like the bell of a stethoscope) before placing them in contact with the patient's body.

"Ice packs" help control swelling after sprains, bruises and muscle strains by "applying cold", i.e. withdrawing heat and causing vasoconstriction. If the ice is wrapped in a wet towel, heat transfer by conduction is facilitated because the wet towel makes good contact with the skin and the ice keeps the wet towel at almost 0 °C until it has all melted.

The body loses little heat through conduction to the air. Conduction to solids and liquids occurs if the bare skin is in contact with something (e.g. water in a swimming pool).

3 Convection

The flow of heated particles from one place to another at a lower temperature is called a convection current and allows heat to be transferred by convection. Clothing our bodies traps air close to the body. This trapped air is heated (by conduction) but is *prevented from forming convection currents* and so prevented from transferring heat away.

Convection may be enhanced by causing the warm air to move away from the body faster – by opening windows to admit breezes or by using electric fans.

The amount of heat lost by convection depends on:

1. The temperature difference between the skin and air and
2. On the area of the exposed skin

The movement of blood around our bodies through our blood vessels is a form of "forced convection". The energy released in one part of our body is transferred to another part, which may be at a lower temperature.

Radiation

All objects emit invisible infrared (IR) electromagnetic radiation due to their temperature.

Radiated energy – $W \propto T^4$.

Infrared thermometers measure this energy.

About half of the human body's heat loss is through radiation.

The amount of heat lost as radiation depends on:

1. The degree of vasodilation of blood vessels in the dermis
2. The temperature difference between the skin and the surrounding objects
3. The surface area of the body
4. Our behaviour

The human sensations of "hot" and "cold" depend on:

1. The local skin temperature
2. The temperature of the object being touched
3. Whether the object is a good or poor conductor of heat

Evaporation of Sweat

The temperature of an object is a measure of the *average* kinetic energy of its particles (hence their speed of movement). The particles have a range of speeds, from low to very high.

Only the fastest (most energetic) water molecules in sweat become vapour (it takes lots of kinetic energy to evaporate). The average kinetic energy of water molecules that have not evaporated is lower (than those that do), so their temperature is lower. These water molecules on the skin are continually gaining energy from blood near the skin's surface. Eventually, these water molecules have enough energy to evaporate. So while evaporation continues, each *evaporating* water molecule removes more than the average amount of kinetic energy (of a water molecule) from the skin, thus leaving the skin at a lower temperature.

If the atmosphere is saturated with water vapour or the humidity is very high, perspiration cannot evaporate, so the effectiveness of this heat loss mechanism is decreased.

Evaporative heat loss is involved when a patient with elevated temperature is sponged with water

Respiratory Heat Loss

We also lose 10–15% of our heat with the air and water vapour we exhale (at 37 °C). The respiratory heat loss from our alveoli may be understood as evaporative loss.

Homeostasis

This refers to the body's automatic tendency to maintain a relatively constant (i.e. oscillating within a narrow range) internal environment (temperature, cardiac output, ion concentrations, pH, hydration, dissolved CO_2 concentration in blood, blood glucose concentration, wastes prevented from accumulating).

The body is in a *dynamic* state of equilibrium; internal conditions change and vary (oscillate) within relatively narrow limits.

Homeostasis returns the body to a healthy state after stressful stimuli by biofeedback mechanisms.

1. RECEPTORS monitor changes in a variable – receive the stimulus.
   - afferent pathway to an integrating centre
2. INTEGRATING CENTRE determines the normal level of the variable – the "set point".
   - efferent pathways to the effector organ
3. EFFECTOR produces a response that moves the variable value back to the set point.

"Negative" feedback-response *opposes* stress.

(For example, the hormonal regulation of blood-sugar level by insulin/glucagon and the nervous regulation of body temperature by shivering, sweating and adaptive behaviour)

If body temp is too low, this is noticed by receptors in hypothalamus and core thermoreceptors. Hypothalamus is the control centre:

- Smooth muscle in blood vessels to vasoconstrict, shivering, release of norepinephrine to increase metabolic rate, behavioural responses (and enhanced thyroxine release in winter).
- Body temp rises (again noticed by hypothalamus and core temperature receptors).
- Hypothalamus turns off heat gain mechanisms.

Positive feedback-response (to infrequent events) *enhances*(!) stress (for example:

- Contractions of childbirth
- Blood clotting with platelets
- Activation of Na^+ channels in the axon hillock when nerve signals are generated
- Ca^{++} combining with diacylglycerol (DAG) to open Ca channels and Ca acting as a "second messenger" in the hormone action).

Disease can be thought of as a disruption of homeostasis or as uncontrolled positive feedback.

For example, in chronic heart failure, the perfusion of peripheral tissues is compromised and coronary blood flow is compromised as well → poor blood flow to the endocardial portion of the heart muscle → subendocardial muscle cells die to be replaced by fibrous tissue → heart becomes weaker → blood flow to the endocardial portion of the heart deteriorates still further…

Regulation of Blood Sugar Levels by Pancreas Hormones

Stimulus:	↑ blood sugar level beyond the set point (after digestion of a meal)
Afferent pathway:	the blood system
Integrating centre:	insulin-secreting cells of the pancreas
Response:	insulin secreting cells ↑ insulin secretion into the blood
Efferent pathway:	the blood system
Effector:	the liver takes up glucose and stores it as glycogen; muscles take up glucose and store it as glycogen; cells take up more glucose.
Result (new stimulus):	blood sugar level ↓ towards the set point (so secretion of insulin ↓)

Stimulus:	↓ blood sugar level below the set point (after a period without eating)
Afferent pathway:	the blood system
Integrating centre:	glucagon-secreting cells of the pancreas
Response:	glucagon secreting cells ↑ glucagon secretion into blood
Efferent pathway:	the blood system
Effector:	the liver breaks down stored glycogen into glucose and releases it into the blood
Result (new stimulus):	blood sugar level ↑ towards the set point (so secretion of glucagon ↓)

4 Thermoregulation Revision: Homework Exercise 4

1. Define what heat is.
2. Define what temperature is.
3. What are the means by which the body can lose heat?
4. Outline the skin's role in temperature regulation.
5. Explain how sweating allows the body to lose heat.
6. Define what is meant by homeostasis.
7. Explain how negative feedback maintains homeostasis.

Lectures 5 and 6: Musculo-Skeletal System

(Comprising: Muscles, Bones, Tendons, Ligaments, Retinaculum, Articulations)

1 Functions of Bones

(a) Support – bones provide an internal framework that supports and anchors all soft organs.
(b) Protection – vertebrae around the spinal cord, skull around the brain, ribs around lungs, patella.
(c) Movement – bones act as levers moved by muscles attached to the bones by tendons.
(d) Blood cell formation (haemopoiesis) – red marrow produces red blood cells, white blood cells and platelets.
(e) Storage – fat stored in internal cavities as yellow marrow; minerals are stored (Ca, P).

2 Bone Naming

There are 206 bones in the (adult) body.

Axial skeleton (80 bones: head, ribs and vertebrae)

- Eight cranial bones (**frontal**, **parietal (two)**, **temporal (two)**, **occipital**, **ethmoid and sphenoid**), 14 facial bones (maxilla (two), zygomatic (two), nasal (two), lacrimal (two), palatine (two), nasal conchae (two), vomer and mandible) and hyoid (one), plus wormians (number varies) and six ossicles in middle ears.
- Twenty-six vertebrae (seven cervical vertebrae, 12 thoracic vertebrae, five lumbar vertebrae, sacrum and coccyx); it has four curves (cervical, thoracic, lumbar and pelvic).

M. Caon, *Lecture Notes, Worksheets, and Exercises for Basic Anatomy and Physiology*, https://doi.org/10.1007/978-3-031-56296-9_5

Thoracic cage: 25 bones (12 pairs of ribs and one sternum)
Appendicular skeleton (126 bones: shoulder girdle and arms/hands, hip girdle and legs/feet)
Pectoral (shoulder) girdles: (two **clavicles** and two **scapulae**)
Upper limbs: 30 bones each (**humerus, radius, ulna, carpals (eight), metacarpals (five)** and **phalanges (14)**)
Pelvic (hip) girdle: two **coxal** (or innominate) bones
Lower limbs: 30 bones each (**femur**, **patella**, **tibia**, **fibula**, **tarsals (seven)**, **metatarsals (five)** and **phalanges (14)**)

Classification of Bones (by Shape)

- Long: humerus, tibia and phalanges
- Short: carpals and tarsals
- Flat: parietal, sternum and ribs
- Irregular: vertebrae and hip bones
- Sesamoid: embedded within a tendon (patella, fabella)
- Sutural: wormian

3 Bone Markings

They fall into two categories:

(i) *Projections* or *processes* (bumps, roughening), which are raised above the bone surface
(ii) *Depressions* or *cavities*, which are indentations in the bone

Projections That Are Sites for Muscle Attachment

Tuberosity	Large rounded projection, may be roughened (e.g. deltoid tuberosity of humerus)
Trochanters	Very large blunt irregular-shaped processs (only on the femur)
Tubercles	Small rounded projections or processes (e.g. greater tubercle of the humerus)
Crest	Narrow, usually prominent, ridge of bone (iliac crest, anterior tibial crest)
Line	Narrow ridge of bone, less prominent than crest (e.g. linea aspera in the femur)
Spine	Sharp, slender, often pointed projection (e.g. posterior-superior iliac spine in the pelvis)

Projections That Help Form Joints (Articulations)

Head	Bony expansion carried on a "neck" (e.g. head of the humerus)
Facet	Smooth, nearly flat articular surface (e.g. facet for the tubercule of the rib)
Condyle	Rounded articular projection (e.g. medial/lateral condyles in the femur)
Ramus	Arm-like bar of bone (e.g. ischial ramus, ramus of mandible)
Fossa	Shallow basin-like depression in a bone, often serving as an articular surface (e.g. olecranon fossa of the humerus, subscapular fossa)

Depressions for Tendons and Openings for Blood Vessels and Nerves

Meatus	Large canal-like passage (e.g. external auditory meatus)
Sinus	Cavity within a bone, filled with air and lined with mucous membrane (e.g. frontal sinus)
Groove	Furrow (e.g. intertubercular groove of the humerus for the tendon of biceps)
Fissure	Narrow slit-like opening (e.g. superior orbital fissure in the sphenoid)
Foramen	Round or oval opening through a bone (e.g. infraorbital foramen, vertebral foramen)

4 Anatomy of the Bone

Gross Anatomy

Long bones: (1) the **diaphysis** (tubular shaft) of compact bone contains a **medullary cavity** and is covered by a **periosteum** (containing osteoblasts – which deposit bone – and osteoclasts – which are multi-nucleate). (2) The **epiphysis** (end) is separated from the shaft by an **epiphysial line/plate**. The interior is spongy bone, and the exterior is compact bone and covered (at articulations) by articular cartilage.

Short, irregular and flat bones: these are thin plates of periosteum-covered compact bone on the outside sandwiching the endosteum-covered spongy bone within.

Histology

Bone is connective tissue, so it has cells (osteocytes, osteoblasts and osteoclasts), a matrix (hydroxyapatite) and fibres within the matrix (collagen).

Two types of bone tissues are distinguished:

1. Compact (hard) bone – e.g. the shaft of a long bone
2. Cancellous (spongy) bone – e.g. the ends of a long bone

Compact bone: the functional units are called Haversian systems or **osteons**. These are concentric cylinders of calcified bone matrix (**lamellae**) between which are scattered **osteocytes** (in the **lacunae**). Lamellae surround Haversian canals (tunnels through the bone which contain blood vessels). The osteocytes in the lacunae are linked to each other and the haversian canal by **canaliculi**. Volkmann's canals run at right angles to the haversian canals and link the periosteum and the medullary cavity.

Spongy bones consist of marrow-filled spaces surrounded by trabeculae (irregularly aligned lamellae and osteocytes interconnected by canaliculi) – no osteons.

Chemical Composition

The bone consists of 60% **inorganic salts** by weight [calcium hydroxyapatite – $Ca_{10}(PO_4)_6(OH_2)$ and $CaCO_3$]

and 40% **organic matrix**, i.e. collagen fibres, glycoproteins (protein + carbohydrate) and cells.

5 Fractures

A fracture is any break in the bone, which may be caused by a sudden injury, fatigue or stress, or pathologic conditions (e.g. neoplasia, osteomalacia, osteomyelitis, osteoporosis).

Sprain = torn ligament or tendon.
Strain = painful overstretch of muscle, tendon or ligament.

Fractures may be classified according to aetiology, location (proximal, distal, midshaft), direction of fracture line (transverse, oblique, spiral), type (open/closed, complete/partial, displaced/non-displaced, greenstick, avulsion) and other descriptors.

Reduction is the restoration of a fractured bone to its normal anatomical position. It is then immobilised.

Union of a fracture occurs when it could already withstand normal stress.

Bone is <u>regenerated</u> rather than healed with a scar.

6 Classification of Joints (Articulations)

1. Functionally
   (a) Immovable joints (**synarthroses**), e.g. sutures between skull bones
   (b) Slightly movable joints (**amphiarthroses**), e.g. between vertebrae
   (c) Freely movable joints (**diarthroses**), e.g. knee, shoulder etc.
2. Structurally (by material between bones)
   (a) **Fibrous joints** – articulating bones are separated by fibrous connective tissue (sutures, tibiofibular (syndesmoses), gomphoses).
   (b) **Cartilaginous joints** – articulating bones have cartilage between them (sternum and rib#1 = synchondrosis, pubic symphysis).
   (c) **Synovial joints** – articulating bones have a fluid-filled space between them, e.g. plane (between carpals), hinge (elbow), pivot (proximal radio-ulnar), ellipsoidal, saddle (thumb carpo-metacarpal) and ball and socket (shoulder).

   <u>Synovial joints</u> are freely movable with the following distinguishing features:

1. Articulating bone surfaces covered by smooth hyaline **cartilage** ~3 mm on the femur and 4 mm on the tibia (which minimises friction).
2. Joint cavity – a potential space filled with lubricating synovial fluid (also provides nutrients, removes wastes, distributes pressure).
3. Joint surrounded by an articular **capsule** composed of two layers: the inner layer of the synovial membrane (secretes synovial fluid) and the outer layer of fibrous connective tissue (often ligaments) attached to the periosteum to limit the range of movement.

4. Reinforcing ligaments – some joints have a fibrocartilaginous pad (meniscus of the knee) to aid shock absorption and improve the fit between bones. Closed fibrous fluid-filled sacs (**bursae**) reduce friction and absorb shock around the joints.

7 Description of Synovial Joint Movement

- Flexion/extension (and hyperextension)
- Abduction/adduction (and circumduction)
- Pronation/supination (of the forearm)
- Inversion/eversion (of the foot)
- Dorsiflexion/plantar flexion (of the foot)
- Rotation

<u>**Lever action**</u>: muscles that pull on bones causing them to move around freely movable joints describe a "**lever action**". Levers are bones that have three forces acting on them:

1. **Fulcrum** (joint)
2. **Effort** (muscle force) and
3. **Load/resistance** (weight of limb)

The arrangement of these three (and the position of the lever in space) determines the <u>class</u> of the lever:

First class: tsri**F F**ulcrum in the middle, e.g. to nod the head to say yes.

Second class: secon**D D**aol (load) in the middle, e.g. to stand on tiptoes.

Third class: drih**T T**roffe (effort) in the middle, e.g. to hold the forearm horizontally then flex it;

this class is common but "inefficient" ➔ the effort exceeds the load that is shifted.

8 Types of Muscles

(a) **Smooth muscle** – not striated, uninucleate, involuntary, not individually named and exists in walls of tubes (e.g. blood vessels, ureters, gut, reproductive tract, bronchioles)

(b) **Cardiac muscle** – cells are "striated", branched, uninucleate, involuntary, and intercalated discs separate adjacent cells

(c) **Skeletal muscle** – striated, multi-nucleate and voluntary (700 muscles, ~40% of body weight)

9 Skeletal Muscles

They produce movement, maintain posture, stabilise joints, generate heat and communicate.

A. Terms

- *Origin*: is a muscle's attachment point to the "stationary" bone
- *Insertion*: is a muscle's attachment point to the "movable" bone
- (Origin and insertion are separated by a joint.)
- *Agonist* (or prime mover): a major muscle, the contraction of which produces specified movement (*In pharmacology, the term agonist-antagonist is used to refer to a drug. Agonist drugs bind to and activate a receptor. An antagonist, on the other hand, is a drug that attenuates the effect of an agonist.*)
- *Antagonist*: opposes or reverses a particular movement
- *Synergist*: aids agonists by promoting the same movement or reducing unnecessary or undesirable movements
- (One muscle may act differently – as an agonist, antagonist and synergist – in different movements.)
- *Fixators*: muscles that immobilise a bone. e.g. scapula is stabilised by (rhomboid major and trapezius) muscles attached to axial skeleton, posture-maintaining muscles are fixators

B. Muscle-Naming Criteria

- Direction of muscle fibres (rectus abdominus, transversus abdominus, external oblique)
- Location in the body (temporalis, tibialis anterior, intercostal, biceps brachii, biceps femoris)
- Relative size (gluteus maximus, gluteus minimus, peroneus longus, peroneus brevis, pectoralis major)
- Number of origins (bi-, tri-, quadriceps)
- Shape (deltoid, trapezius, teres)
- Origin and insertion (sternocleidomastoid, pubococcygeus)
- Action (adductor longus, extensor digitorum longus, flexor carpi radialis, supinator)
- Whimsy (sartorius, gracilis, soleus)

C. Superficial Muscles

You should know (or be able to find) the positions, actions, approximate insertions and origin of the following:

(a) Head (mastication): temporalis, masseter, buccinator, tongue
(b) Neck, upper chest: sternocleidomastoid, deltoid, pectoralis major, serratus anterior
(c) Abdomen: external obliques, rectus abdominus
(d) Upper arm: biceps brachii, brachialis, triceps brachii

(e) Lower arm: forearm flexors (brachioradialis, flexor carpi radialis, palmaris longus, flexor carpi ulnaris), forearm extensors (extensor carpi ulnaris, extensor digiti minimi, extensor digitorum, extensor carpi radialis brevis)
(f) Back: trapezius, latissimus dorsi
(g) Buttocks: gluteus maximus and medius
(h) Thigh: hamstrings (biceps femoris, semitendinosus, semimembranosus), quadriceps femoris (vastus lateralis, vastus intermedius, vastus medialis, rectus femoris)
(i) Leg: gastrocnemius, soleus (together = triceps surae), tibialis anterior.

The deep thoracic muscles promote movement for breathing:

(a) **External intercostals** – a more superficial layer that lifts the rib cage and increases thoracic volume to allow inspiration
(b) **Internal intercostals** – a deeper layer that aids in forced expiration
(c) **Diaphragm** – the most important muscle in inspiration (accessory muscles for forced inspiration = sternocleidomastoid, scalenes, pectoralis minor)

D. <u>Muscle-Bone Attachment</u>

Muscles attach to the bone or cartilage via (rope-like) **tendons** or a (flat broad) **aponeurosis**.

10 Histology of Muscle

The skeletal muscle cell (= a muscle fibre) is a huge multi-nucleate cell (a syncytium), 3cm long on average, can be up to 30 cm long, and 10–100 microns diameter.

Plasma membrane = **sarco**lemma; it has invaginations (T-tubules).

Cytoplasm = **sarco**plasm (filled with myofibrils of 1–2 microns in diameter).

Endoplasmic reticulum = **sarco**plasmic reticulum.

The endomysium (connective tissue) surrounds individual muscle cells and contains capillaries, nerves and myosatellite cells (stem cells that repair damaged muscle cells).

Myosin is the protein of thick myofilaments.
Actin (and tropomyosin and troponin) is the protein of thin myofilaments.
Thick myofilament is composed of myosin (15nm).
Thin myofilament is composed of actin (6nm).
Sarcomere is a bundle of interdigitated thick and thin myofibrils.
Myofibril is a long line of sarcomeres joined end to end.
Muscle fibre (a cell containing many myofibrils).
Fasciculus (contains many muscle fibres).
Muscle (contains many fasciculi).

(A muscle contains many fasciculi, a fasciculus contains many muscle fibres (cells), a muscle fibre is composed of many myofibrils, myofibrils are many sarcomeres joined end to end, a sarcomere is a bundle of thick and thin myofilaments, thick myofilaments are made of the protein myosin, while thin myofilaments are made of the proteins actin, tropomyosin and troponin.)

Neuromuscular junction, the meeting point of the motor end plate on the sarcolemma of a muscle cell and the axon terminal of a motor nerve cell, is a gap crossed by ACh (acetylcholine).

Sarcomere is a length of myofibril – between two "Z-lines" – in which we identify five sections using the letters: I, A, H, A and I defined below.

As thick and thin myofilaments interdigitate, the sarcomere shortens, which in turn shortens the myofibril; hence, the muscle fibre contracts.

The Z-line is attached to the I-band (thin myofilaments only).

The A-band consists of thick and thin myofilaments together.

The "H-zone" (which contains only thick filaments when relaxed) exists within the A-band. Thin filaments move into the H-zone during contraction.

The I-band is attached to the Z-line.

11 Mechanism of Contraction

Consider the events between depolarisation and contraction: one cycle of the sliding filament mechanism

1. ACh from the axon terminal causes an action potential (i.e. depolarisation) to propagate along the sarcolemma and into the T-tubules, which pass through the sarcoplasmic reticulum (SR). (ACh is then destroyed.)
2. Depolarisation causes the release of Ca^{2+} (from the terminal cisternae of the sarcoplasmic reticulum).
3. Ca^{2+} diffuses through the sarcoplasm and binds to the troponin. The troponin changes shape and pulls the tropomyosin away from the actin molecule.
4. The energised myosin head (with adenosine di-phosphate (ADP) + inorganic phosphate (P_i)) from the thick myofilament is able to engage in (the now exposed) binding site on the actin molecule; that is, a "cross-bridge" is formed.
5. The engagement of the cross-bridge causes the myosin molecule to change shape, resulting in a "power stroke" (a thick filament sliding over the thin filament until they overlap; the muscle fibre contracts, i.e. the muscle *fibre* becomes shorter than before, but the *myofilaments* are of the same length).
6. ADP + P_i detach from the head of the myosin cross-bridge. This allows the adenosine tri-phosphate (ATP) to attach to the place vacated by ADP.
7. The binding of ATP causes the myosin head to disengage from the binding site on the actin molecule (thin filament).
8. The cross-bridge retreats from the thin filament, and ATP hydrolyses into ADP + P_i + energy. Energy is stored in the myosin head – it is "energised".

   Note: ATP = (ADP + P_i + energy); this energy is stored in the cross-bridge (P_i = HPO_4^{2-}).

12 Energy for Contraction

ATP is an energy source. Its rate of production must equal its rate of use since very little is stored in the muscle – enough for only a few seconds of vigorous activity.

Three Methods of Production of ATP

1. *Direct phosphorylation of ADP by creatine phosphate (CP)*: CP breaks down to creatine and energy, and the energy is used to form ATP from ADP. This mechanism provides enough energy for about 15 s of vigorous activity.

$$CP + \text{energy} + ADP \rightarrow ATP + \text{C}.$$

2. *Anaerobic glycolysis and the production of lactic acid*: glucose from the blood (which enters fibres through facilitated diffusion) and from the breakdown of muscle glycogen is converted to pyruvic acid and energy for ATP; then (in the absence of oxygen) pyruvic acid is converted to lactic acid and energy for ATP.

   This mechanism provides enough energy for about 30 to 100 s of vigorous activity – such as 400 m sprints.

$$\text{Glucose} + 2\,ATP \rightarrow 2\,\text{pyruvic acid} + 4\,ATP, \ \text{then pyruvic acid} \rightarrow \text{lactic acid}.$$

3. *Aerobic respiration*: in the presence of O_2, pyruvic acid enters the mitochondria, and a slow series of reactions called oxidative phosphorylation occurs.

$$\text{Pyruvic acid} + \text{O}_2 \rightarrow 18\,ATP + CO_2 + \text{heat}.$$

   This mechanism provides enough energy for about 40 min of vigorous activity – such as jogging.

13 Intramuscular (IM) Injection Sites

(i) Gluteus medius (avoids sciatic nerve)
(ii) Vastus lateralis
(iii) Deltoid

All are easily accessible, thick muscles, with a large number of muscle fibres, extensive fascia ➔ large surface area for absorption and an extensive blood supply.

IM injection is used for the prompt absorption of larger doses than can be given subcutaneously (or too irritating for SC).

Drugs with low solubility in water can be administered IM. No cell membrane to cross to be absorbed into capillaries (pass through spaces between cells of capillary wall). May be painful.

14 Pelvic Floor (Pelvic Diaphragm) Muscles

The "pelvic floor" separates the pelvic cavity from the perineum.

Left and right levator ani muscles consist of:

Iliococcygeus
Pubococcygeus
Puborectalis - which may be part of the external anal sphincter (sphincter ani externis). The coccygeous (left and right) are frequently tendinous rather than muscular. The vaginal orifice, urethral orifice & anal orifice, all penetrate the pelvic floor.

(Why does the knee have such a complicated arrangement of ligaments?)

Tendons and Ligaments That Cross the Knee Joint

Ligaments
- Quadriceps femoris tendon-patellar ligament
- Fibular collateral ligament
- Tibial collateral ligament
- Oblique popliteal ligament
- Anterior cruciate ligament
- Posterior cruciate ligament

Tendons
- Gracilis tendon
- Sartorius tendon
- Semitendinosus tendon
- Semimembranosus tendon
- Biceps femoris tendon
- Iliotibial tract (from gluteus the maximus and tensor fasciae latae)
- Gastrocnemius tendon
- Popliteus tendon
- Plantaris tendon

(**Osteoarthritis** is a chronic disease characterised by pain, biochemical and enzymatic changes, cartilage fragmentation and loss, osteophyte formation and bony sclerosis. It is the most prevalent and costly joint disease in older adults. In the Western world, at least 10% of the population has symptoms of this disease. Around ? 50% will require (?) treatment with drugs. Treat this with strontium ranelate, which stimulates bone formation and inhibits bone resorption. Also, it strongly stimulates human cartilage formation in vitro.

Rheumatoid arthritis is an autoimmune disease characterised by chronic inflammation of the synovial membrane leading to deformity, loss of function, pain, swelling and, ultimately, joint destruction.

Psoriatic arthritis is associated with the skin condition psoriasis.)

15 Additional Information

Osteoporosis

Osteoporosis = low bone mass ➔ vertebral fractures (back pain) and hip fractures (very debilitating).

Osteoporosis = bone mineral density (BMD) t-score ≤ -2.5 (?means what?).

Vitamin D deficiency = serum 25 OHD level (25-hydroxyvitamin D) < 9 ng/mL ➔ osteomalacia.

Vitamin D insufficiency = serum 25 OHD level (25-hydroxyvitamin D) < 15 ng/mL.

Osteoporosis takes 1000–1500 mg Ca and 400–1200 IU of vitamin D daily.

Serum Ca ref values are 2.08–2.60 mmol/l.

Zoledronic acid is a third-gen bisphosphonate drug (phosphorus-carbon-phosphorous bond is resistant to phosphatases and binds to calcified bone matrix and inhibits osteoclastic bone resorption). Intravenous (IV) injection 4 mg once per year increases bone mass.

The majority of vitamin D is produced in the skin when ultraviolet B (UVB) photons stimulate the conversion of 7-dehydrocholesterol to pre-vitamin D. Pre-vitamin D is then spontaneously converted to vitamin D_3 and sequentially converted to 25-hydroxyvitamin D_3 in the liver and to the active hormone 1,25-dihydroxyvitamin D_3 in the kidney. Vitamin D_2, a yeast-and-plant-derived vitamin D analogue, is converted by the same enzymes, and its metabolites have approximately the same biological activity as those of vitamin D_3. The principal action of 1,25-dihydroxyvitamin D_3 is to increase the intestinal absorption of both calcium and phosphate as well as to regulate bone turnover. Circulating 1,25-dihydroxyvitamin D_3 decreases serum parathyroid hormone (PTH) by indirectly increasing serum Ca levels (thus decreasing PTH release) and by directly decreasing parathyroid gland activity.

(USA data) Cardiovascular disease is the leading cause of death in women (exceeds the next 16 causes of death combined and more than twice the number of all cancer deaths combined).

Skin cancer incidence is > breast cancer incidence (breast cancer is one in every three cancer diagnoses in the USA).

Lung cancer mortality > breast cancer mortality.

Paget's Disease

A chronic bone disorder is characterised by localised areas of accelerated bone remodelling leading to structurally flawed bone ➔ bone deformity, pain, fractures and deafness. It can affect any bone but mainly the pelvis, femur, skull, tibia, vertebrae, clavicle and humerus.

It has a prevalence in 4% of people aged over 55 (a 1978 WA radiological study). It is treated with drugs like biphosphonates, e.g. zoledronic acid.

Raised serum alkaline phosphotase is one of the main indicators of Paget's disease.

16 Musculo-Skeletal System Revision: Homework Exercise 5

1. List the functions of the skeletal system. What are the two major divisions of the skeletal system?
2. How does compact bone differ from cancellous bone?
3. Describe the histology of bone (and name the structures).
4. Characterise the following seven types of joints (symphyses, syndesmoses, synchondroses, gomphoses, sutures, hinge, and ball and socket) according to:

   (a) Structure, as one of fibrous, cartilaginous or synovial.

   Fibrous joints:
   Cartilaginous joints:
   Synovial joints:

   (b) Function, as one of syn-, amphi-, or diarthroses.

   Synarthroses:
   Amphiarthroses:
   Diarthroses:

5. Give an example of each of the following types of bones: long, short, flat, irregular and sesamoid.

   Long:
   Short:
   Flat:
   Irregular:
   Sesamoid:

6. Perform the following movement with your left upper extremity: with the thumb and third fourth fingers fully flexed and the first and second fingers fully extended, circumduct the arm in a flexed position with the lower arm pronated.
7. Compare AND contrast the anatomy of the knee joint and elbow joint. In what way(s) does(do) their structure reflect their function?
8. Find an example of a superficial muscle that is descriptively named according to each of the eight types of descriptors: direction of muscle fibres, location in the body, relative size, number of origins, shape, origin and insertion, action and whimsy.

   Direction of muscle fibres:
   Location in the body:
   Relative size:
   Number of origins:
   Shape:
   Origin and insertion:
   Action:
   Whimsy:

9. Choose six muscles (different from those in Q 8), one from each of the lower arm and leg, upper arm and leg, and front and back of the body, and list their origin, insertion and action.

   Forearm:
   Leg:
   Arm:
   Thigh:
   Anterior of the body:
   Posterior of the body:

10. Describe the relationship between the three proteins actin, troponin and tropomyosin in thin myofilaments.
11. In which directions do the muscle fibres in the external obliques and rectus abdominus muscles lie?
12. Comment on the relative size of the gluteus maximus and the gluteus medius.
13. Which muscles extend the spine, and which muscles extend the arm?
14. What actions do the following muscles perform?

    (a) Pronator quadratus
    (b) Adductor magnus
    (c) Extensor digitorum
    (d) Sartorius

15. Over which bones do the extensor carpi ulnaris and the extensor carpi radialis brevis lie?
16. What part of the name biceps femoris indicates its location?
17. How many origins do the triceps brachii have, and where are they?
18. What is the origin and insertion of the sternocleidomastoid muscle?
19. What broad general shapes do the (a) rhomboid major muscle and (b) deltoid muscle resemble?
20. What action is performed by the masseter?
21. Under which gluteal muscle does the sciatic nerve lie?
22. List the muscles that comprise the quadriceps group. Where are they?
23. List the muscles that comprise the hamstrings group.
24. Where are the locations of the triceps brachii and the triceps surae?
25. Give five reasons why the deltoid, vastus lateralis and gluteus medius muscles are preferentially chosen as the sites for intramuscular injection.

Lectures 7 and 8: Digestive System

1 Gastrointestinal (Alimentary) System and Accessory Digestive Organs

GIT organs	Accessory organs
Mouth	Teeth and tongue
Pharynx	Salivary glands
Oesophagus	* Parotids submandibular and
Stomach	* Sublingual glands
Small intestine	
Duodenum	Pancreas
Jejunum	Liver and gall bladder
Ileum	
Large intestine	
Caecum and appendix	
Ascending colon	
Transverse colon	
Descending colon	
Sigmoid colon	
Rectum	
Anus	

2 Digestive Processes

Digestion (disassembly) is necessary to reduce the very large molecules (polymers) ingested as food to particles small enough to pass into the cells lining the gut.

M. Caon, *Lecture Notes, Worksheets, and Exercises for Basic Anatomy and Physiology*, https://doi.org/10.1007/978-3-031-56296-9_6

(i) Ingestion
(ii) Propulsion (swallowing and peristalsis)
(iii) Mechanical digestion
(iv) Physically preparing food for digestion
(v) Chewing, mixing with saliva, churning in the stomach and *segmentation* (mixes food with digestive enzymes and moves food over the intestinal wall)
(vi) Chemical digestion
(vii) Hydrolysation of complex food molecules into monomers via enzymes
(viii) Absorption
(ix) Passage of end products of digestion into the blood or lymph after entering mucosal cells by active or passive transport processes
(x) Defecation
(xi) Elimination of indigestible substances (faeces)

3 Histology of the Alimentary Canal

From the oesophagus to the anal canal, the wall of the alimentary canal is made up of the same four layers (called *tunics*) 2–5 mm thick:

1. *Mucosa (Includes the Lamina Propria)*
    - *Secretes* mucus, digestive enzymes and hormones
    - *Absorbs* end products of digestion
    - (Lymphoid nodules) *protects* against infectious disease and self-digestion
    - (Muscularis mucosae) *wiggles* the epithelial folds
2. *Sub-mucosa*
    - Dense connective tissue containing blood vessels, lymphatics (transport products of digestion and combat bacteria) and nerve plexus (innervate gut).
3. *Muscularis Externa*
    - Gut motility: peristalsis (propulsion) and segmentation (mixing)
    - Inner circular muscle layer (→ sphincters), outer longitudinal muscle layer
4. *Serosa (= Visceral Peritoneum)*
    - Areolar connective tissue, which suspends the gut's abdominal organs (however, the oesophagus has adventitia – a fibrous connective tissue)
    - Visceral peritoneum attached to the parietal peritoneum by double-sheeted mesenteries (greater omentum, lesser omentum, falciform ligament, mesentery proper, mesocolon), which attach the gut to the adjacent structures.
    - Cardiac (gastro-oesophageal) sphincter, pyloric sphincter, ilio-cecal valve, internal anal sphincter, external anal sphincter

4 Functional Anatomy

Mouth

- Food ingestion
- Mechanical digestion, mixing (teeth, tongue) to increase the surface area of food for enzyme access.
- Chemical digestion (salivary glands produce saliva, salivary amylase acts on starch, lysozyme – a proteolytic enzyme that acts on bacteria and protein)
- Propulsion of food into the pharynx

Pharynx

Has two layers of *skeletal muscle* to propel food into the oesophagus (deglutition).

Oesophagus

A muscular tube that conducts food to the stomach

Stomach

The surface epithelium of the stomach mucosa is a simple columnar epithelium composed of mucous cells.

Gastric glands below the epithelium secrete "gastric juice":

1. Mucous cells secrete *mucus*.
2. Parietal cells secrete H^+Cl^- – H^+ from H_2O via carbonic anhydrase (required to activate pepsinogen and kill bacteria and unfold (denature) protein molecules) and *intrinsic factor* (required for the absorption of vitamin B_{12} in the small intestine (SI))
3. Zymogenic (chief) cells secrete *pepsinogen*.
4. Enteroendocrine G cells secrete the *hormone* gastrin, D cells secrete somatostatin and P/D1 cells secrete ghrelin (stimulates hunger) and obestatin (induces satiety).

The **mucosal barrier** prevents the stomach from digesting itself (bicarbonate in the mucus reacts with HCl → CO_2, stomach epithelial cells joined by "tight" junctions).

Food Storage

The sub-mucosa has elastic fibres enabling the stomach to regain shape after storing a large meal.

The muscularis externa has three layers of smooth muscle. They relax to allow the stomach to expand and stretch without increasing their tension.

Mechanical Digestion

Mixing waves (3/min) churn contents to mix chyme and gastric juice. About 3 ml at a time of liquid chyme squirts through the pyloric sphincter into the duodenum.

Chemical Digestion

Pepsinogen + HCl → pepsin.

Pepsin digests (hydrolyses) protein to polypeptides.

Regulation of Gastric Secretion (Three Phases)

1. Cephalic (Reflex) Phase

   Thought, sight, smell and taste of food – a conditioned reflex that increases gastric secretion (occurs before food enters the stomach)

2. Gastric Phase

   Stomach distension increases secretion.

   Peptides (partially digested proteins) and caffeine stimulate the release of **gastrin** (a hormone that stimulates the secretion of hydrochloric acid (HCl)).

   Also, a rise in pH level stimulates the release of gastrin.

3. Intestinal Phase
   (a) Excitatory: food entering the duodenum stimulates the release of **intestinal gastrin**, a hormone that stimulates gastric secretion (briefly).
   (b) Inhibitory: the distension of the duodenum and the presence of acidic chyme, fats and partially digested food trigger the **enterogastric reflex**. This produces the tightening of the pyloric sphincter and causes a decrease in gastric secretion.

In addition, intestinal hormones are released (**secretin**, **cholecystokinin** (CCK), vasoactive intestinal peptide), inhibiting gastric secretion.

CCK and secretin also stimulate the release of bile and pancreatic juice.

Small-Diameter Intestine (Duodenum, Jejunum and Ileum)

It is the major digestive and absorptive organ, about 2 m long (during life, but longer when muscle tone is lost and a cadaver is dissected!) and ~2.5 cm in diameter.

The internal absorptive surface area increased beyond that available to a smooth tube by:

- Plicae circularis (circular folds 1 cm deep)
- Villi (finger-like projections 1 mm long)
- Microvilli (tiny projections of the plasma membrane of mucosal cells), the "brush border"

The mucosa consists of:

- Simple columnar epithelium (absorptive cells)
- Goblet cells (secrete mucus)
- Entero-endocrine cells (secrete secretin, CCK)
- Paneth cells (secrete lysozyme, phagocytic)
- Intestinal crypts, which secrete "intestinal juice"

The sub-mucosa consists of areolar connective tissue and contains:

- Lymphoid nodules (called Peyer's patches) to combat bacteria
- Mucus-secreting duodenal glands (bicarbonate-rich mucus to neutralise stomach acid)

Large-Diameter Intestine (Cecum, Appendix, Colon, Rectum and Anal canal)

The large intestine (LI) is about 2 m long, functions to absorb water (H_2O) from indigestible food residue and expels faeces (defecation).

Electrolytes (Na^+ & Cl^-) are absorbed.

Some bacteria are not killed by the action of lysozyme, defensins, HCl and protein-digesting enzymes. They are the bacterial flora of the LI. They ferment some indigestible carbohydrates (cellulose) and produce gas (flatus).

The flora also synthesises B complex vitamins (thiamine, riboflavin, B_{12}) and vitamin K2 (required by the liver to synthesise clotting proteins).

5 Chemical Digestion in the Small Intestine

Chemical digestion involves the *hydrolysis* of carbohydrates, peptides and lipids.

Hydrolysis is the splitting of a large molecule into two smaller molecules by the addition of a (or a few) water molecule(s). The addition is *catalysed* by enzymes.

$$\underset{\text{Sucrose}}{C_{12}H_{22}O_{11}} + H_2O \rightarrow \underset{\text{Glucose}}{C_6H_{12}O_6} + \underset{\text{Fructose}}{C_6H_{12}O_6}$$

Enzymes are from the "brush border" of the SI and from the pancreas.

Carbohydrates are hydrolysed with the aid of dextrinase, glucoamylase, lactase, maltase, sucrase (from the brush border) and pancreatic amylase.
The products are monosaccharides.

Proteins are hydrolysed with the aid of carboxypeptidase, aminopeptidase and dipeptidase (from the brush border) and trypsin and chymotrypsin (from the pancreas).
The products are amino acids and small peptides.

Lipids (triglycerides and their hydrolysis products) are insoluble in water. The fat globules that they form are emulsified by the detergent action of bile salts. The emulsified fatty droplets are hydrolysed with the aid of pancreatic lipases (lingual lipase and gastric lipase also contribute to the digestion of fats). Phospholipids are hydrolysed by phospholipidases; cholesterol does not need to be digested for absorption.
The products are free fatty acids (FFAs) and monoglycerols.

Nucleic acids (deoxyribonucleic acid (DNA) and ribonucleic acid (RNA)) are hydrolysed by pancreatic nucleases into nucleotides and then hydrolysed by brush border enzymes (nucleosidases and phosphatases) into their free bases, pentose sugars and phosphate (PO_4) ions.

6 Absorption in the Small Intestine

Virtually, all foodstuffs, 80% of electrolytes (iron in the duodenum) and most of the water are absorbed in the small intestine (before the ileum). Endothelial cells are joined by "tight junctions" → molecules must pass through their plasma membrane (cannot pass between them). Is this passive or active transport?

Monosaccharides and amino acids are actively transported into the epithelial cells, pass through them unchanged and enter the interstitial fluid before entering the blood capillaries (to be carried to the liver by the hepatic portal vein).

FFA and monoglycerols enter the epithelial cells by diffusion through the PlaM (from micelles) and are reconstituted into triglycerides. They then combine with phospholipids and cholesterol and are coated with protein to form water-soluble **chylomicrons**. Chylomicrons enter the lymphatic system (via lacteals) and eventually the bloodstream via the thoracic duct.

Bile is absorbed from the ileum and recycled.

(Vitamin B_{12} + intrinsic factor complex are taken up by endocytosis from the lower end of the ileum.)

7 Accessory Digestive Organs

Chemical digestion in the SI depends on the function of the following three accessory organs.

Pancreas

It lies mostly retroperitoneal and dorsal to the stomach.

It is connected to the duodenum via the "main pancreatic duct". The *exocrine* function of the pancreas is accomplished by clusters of secretory cells called **acini**. These produce alkaline pancreatic juice (enzymes and bicarbonate ions).

Pancreatic Juice Enzymes

- Amylase (digests carbohydrates)
- Lipase (digests fats and oils)
- Nuclease (digests nucleic acids)
- <u>*Inactive*</u> procarboxypeptidase, chymotrypsinogen, trypsinogen and proelastase

(*Trypsinogen* is activated to *trypsin* in the duodenum by enteropeptidase and in turn activates *carboxypeptidase*, *elastase* and *chymotrypsin* so the pancreas does not digest itself.)

The pancreas is stimulated to release enzymes by **cholecystokinin** (CCK) (a local hormone from the duodenal mucosa), which is released when protein and fats enter the duodenum, and by **secretin** (a hormone from duodenal mucosa), released when HCl is present in the duodenum.

Bicarbonate Ions (HCO_3^-)

They cause pancreatic juice alkaline (pH 7.1–8.2) to raise the pH of acidic chyme from the stomach.

$$H^+ + Cl^- + Na^+ + HCO_3^- \rightarrow Na^+ + Cl^- + H_2CO_3.$$

Carbonic acid dissociates into carbon dioxide (CO_2) and water.

Liver

This is the large (1.4 kg) organ in the right upper abdomen under the diaphragm inside the ribcage. It filters monosaccharides and amino acids out of the blood from the gastrointestinal tract (GIT) and processes them.

The functional unit of the liver is the (microscopic) six-sided **lobule** consisting of **hepatocytes**.

Hepatocytes are arranged in lines that radiate from a central vein. Between lines of hepatocytes are blood-filled **sinusoids** (without basal lamina) containing **Kupffer cells** (macrophages), which remove debris such as bacteria and worn-out blood cells.

Hepatocytes secrete bile into tiny canals (canaliculi), which drain into bile ducts (at the edge of a lobule), which drain into the common hepatic duct, which joins with the cystic duct (from the gall bladder (GB)) to become a bile duct.

Blood Supply of a Lobule

1. Hepatic artery (carrying oxygenated blood)
2. Hepatic portal vein (deoxygenated blood with nutrients absorbed from the digestive tract)

Blood from the hepatic artery and hepatic portal vein enters the portal arteriole and portal venule (respectively), which are alongside the bile ducts in each lobule. The arterial and venous blood mix in the sinusoids and drain into the central vein of each lobule.

This mixed blood enters the hepatic vein, which drains into the inferior vena cava and enters the right atrium of the heart.

Functions of Liver Hepatocytes 1. ***Carbohydrate Metabolism***

It maintains blood glucose levels at 3.3–6.1 mmol/l (glucose buffer function)

(a) If blood glucose is low (fasting)

- Glycogen stored in the liver is converted to glucose and released into the blood (glycogenolysis).
- Amino acids are deaminated and converted to glucose (gluconeogenesis).
- Other sugars (galactose and fructose) are converted to glucose.
- (**Deamination** = removal of an amine group – NH –from a molecule – it produces (toxic) ammonia.
- **Gluconeogenesis** = conversion of non-carbohydrate molecules to glucose.)

(b) If blood glucose is high (after meal)

- Glucose is absorbed and converted to glycogen or triglycerides and stored in the liver and adipose tissue.

2. ***Lipid Metabolism***

- Stores triglycerides
- Converts fatty acids to acetyl CoA and then to ketones
- Synthesises lipoproteins to transport fatty acids, triglycerides and cholesterol
- Synthesises cholesterol from acetyl CoA for use in bile salts
- (Converts carbohydrates and proteins to fats, which are then transported in lipoproteins to adipose tissue)

3. ***Protein Metabolism***

- Synthesises plasma proteins (albumin, clotting proteins, transport proteins) and enters sinusoids easily as there is no basal lamina
- Deaminates amino acids to form keto acids, which are then used for adenosine triphosphate (ATP) or glucose synthesis or converted to fat
- Converts toxic NH_3 from deamination into (less harmful) **urea** for excretion in urine
- Transamination – forms non-essential amino acids from the parent keto acid

4. ***Removal of Drugs and Hormones and Antibodies***

- Metabolises blood-borne hormones to forms that may be excreted in urine
- Detoxifies alcohol
- Metabolises drugs (via cytochrome P450 isozymes) – sulphonamides, penicillin, ampicillin, erythromycin (inactivates them for excretion in the bile or alters their activity)
- (e.g. converts *capecitabine* – an oral anti-cancer pro-drug – into a precursor of *5-fluorouracil*)

5. ***Excretion of Bilirubin***

(In the spleen, red blood cells (RBC) break down, haeme is split from the haemoglobin, iron (Fe) is salvaged and stored and the remaining haeme fragment is degraded to bilirubin.) The liver picks up bilirubin and secretes it in bile.

(Bilirubin → urobilinogen (by bacteria in the gut) → urobilin and stercobilin (a yellow-brown pigment, hence faecal colour))

(The bilirubin in blood should be < 20 μmol/l. If >34 μmol/l, then there is impaired liver function.)

6. ***Storage of Fat-Soluble Vitamins (A, B_{12}, D, E, K)***

1–2 year supply of A

1–4 months supply of D and B_{12}

7. ***Storage of Minerals***

Fe (as ferritin) and copper (Cu)

8. ***Phagocytosis***

Kupffer's cells (in the sinusoids of liver lobules) remove bacteria originating from the intestines, as well as worn-out RBC and white blood cells (WBC).

9. ***Activation of Vitamin D***

   It modifies cholecalciferol (pro-vitamin D_3) prior to its activation in the kidney – D is needed for the absorption of calcium (Ca) from the GI tract.

10. ***Synthesis of Bile Salts***

    Bile salts are needed in the small intestine for the emulsification of fats.

11. ***Storage of Blood***

    The liver normally holds ~ 10% (450 ml) of the body's blood. It can expand to hold up to a litre more, so it stores or supplies blood as required.

12. ***Production of the "Pro-hormone" Angiotensinogen***

    (+ renin ➔ angiotensin I) (+ACE ➔ angiotensin II).

    It converts the hormone cortisol to cortisone.

13. ***Lactic Acid (LA) Removal and Recycling***

    LA (produced by **an**aerobic muscle metabolism of pyruvic acid) is converted back to pyruvic acid, then to glucose (gluconeogenesis).

14. ***Synthesis of Plasma Proteins***

    Examples of plasma proteins are albumin, globulin (e.g. angiotensinogen), prothrombin, fibrinogen and transferrin.

    The liver can regenerate! The turnover of cells is gradual.

    Hepatitis is an inflammatory disease of the liver.

    Prolonged inflammation, e.g. due to chronic alcoholism, may cause cirrhosis (hepatocytes replaced by connective tissue).

Gall Bladder

It is a muscular sac located on the posterior surface of the liver, which stores and concentrates bile (by absorbing water and ions).

When acidic fatty acids enter the SI, **CCK** is released. This causes the smooth muscle of the GB to contract and the hepatopancreatic sphincter to relax; hence, bile enters the duodenum.

Bile consists of H_2O, bile salts (cholic acid, chenodeoxycholic acid, glycolic acid, taurocholic acid), bile pigments (bilirubin and biliverdin), cholesterol, lecithin (a phospholipid), fatty acids and ions (Na^+, K^+, Ca^{++}, Cl^-, HCO_3^-).

Note: Bile salts have a hydrophobic tail (which dissolves in fats) and an ionic head (which dissolves in water).

Vagus Nerve

- It is the longest of the cranial nerves.
- It can be thought of as the "spinal cord" of the autonomic nervous system (NS).

- Together with three sacral nerves, the vagus is the primary pathway for parasympathetic NS communication to and from the brain.
- The vagus regulates much of the digestive system:
  - Stomach

    Post-prandial gastric volume (accommodation)
    Gastric contractions (food processing)
    Acid secretion
    Gastric emptying

  - Pancreas and gall bladder

    Pancreatic exocrine secretion
    Gall bladder emptying
    Bile flow

  - Small intestine

    Intestinal motility
    Intestinal content transit

Absorption of Nutrients from the Small Intestine
Digestion is the chemical and mechanical breakdown of food into smaller units that can be taken across the intestinal epithelium into the body.

MONOSACCHARIDES are transported into epithelial cells by protein carriers and then through facilitated diffusion into capillary blood (fructose moves entirely through facilitated diffusion).

As well as specific AMINO ACIDS, short chains of two or three amino acids (dipeptides and tripeptides) can be actively transported into epithelial cells. Di- and tripeptides are digested in epithelial cells into amino acids before entering capillary blood by diffusion.

8 Additional Information

MONOGLYCERIDES and FREE FATTY ACIDS (FFAs) become associated with bile salts and lecithin to form **MICELLES** (also included are cholesterol and fat-soluble vitamins).

- At the plasma membrane of epithelial cells, monoglycerides, FFA, cholesterol, and fat-soluble vitamins leave micelles and diffuse through the lipid plasma membrane.
- Inside epithelial cells, FFA and monoglycerides are reconstituted to triglycerides.
- These combine with phospholipids and cholesterol and gain a protein skin to form **CHYLOMICRONS**.

Chylomicrons are processed by a Golgi apparatus for extrusion from the cell to pass into lacteal ➔ the lymphatic system ➔ venous blood via the thoracic duct.

NUCLEIC ACIDS: the pentose sugars, nitrogenous bases and phosphate ions are transported actively across the epithelium by special carriers.

Vitamin B_{12} (cobalamin) is absorbed when complexed with *intrinsic factor*, which is secreted by the stomach.

Mineral absorption (magnesium (Mg), Fe, Ca) and ions (PO_4, sulphate (SO_4)) occur by active transport (the absorption of Fe and Ca is linked to their concentration in the body); water osmotically follows electrolytes.

Toxic Nitrogenous Wastes

Ammonia – from the deamination of amino acids
Urea – from protein catabolism (3–8 mmol/L)
Uric acid – from the metabolism of nucleic acids
Creatinine – from creatine phosphate (0.06–0.13 mmol/L)

HCL Production in Parietal Cells of the Stomach Wall

H_2O ➔ $H^+ + OH^-$
(H_3O^+ transported quickly into the stomach lumen in exchange for K^+)
$OH^- + CO_2$ ➔ HCO_3^- via carbonic anhydrase
(HCO_3^- transported into interstitial fluid in exchange for Cl^-, Cl^- leaks into the stomach)
This increase in HCO_3^- in stomach blood supply raises blood pH slightly (alkaline tide).
➔ Vomiting causes the loss of Cl^- and K^+ as well as acid.
pH parietal cells = 7.2
pH stomach = 1 or 2, so the [H_3O^+] gradient is large; the stomach has 10,000 times the acid concentration of parietal cells.

Portal Hypertension

Cirrhosis of the liver accounts for 90% of cases of portal hypertension. Bands of fibrous tissue distort liver sinuses and increase resistance to portal venous flow. The resulting increased pressure causes large collateral channels to develop between the portal and systemic veins that supply the lower rectum and oesophagus (and in the umbilical veins in people in which these have not been obliterated).

This gives rise to haemorrhoids and dilated veins around the umbilicus.

The most important collateral channels are those that divert blood to the coronary (gastric) **veins** and form thin-walled varicosities in the sub-mucosa of the oesophagus. They are subject to rupture, with massive and sometimes fatal haemorrhage.

Pancreatic cancer is the fourth leading cause of cancer deaths in the USA. Incidence in Europe $11.4/10^5$ in males, mortality rate $11.3/10^5$. pt with advanced pancreatic cancer median survival without therapy = 15 weeks, with chemo and radiotherapy, med survival = 47 weeks.

Gastric cancer incidence in Japan is $69/10^5$ males (lower in females and lower in Western countries).

Bowel cancer is the most common internal cancer affecting Australians and the second-highest cause of cancer-related deaths, lagging behind lung cancer.

Overweight and Obesity
Excess energy not required by the body is stored as sub-cutaneous fat, and the amount stored in this form varies greatly between individuals. The stored fat is usually measured as a percentage of total body mass. A percentage body fat (%BF) of 15% or less is **desirable** for men, and 25% or less for women is desirable. Women generally have a greater fat store than men to support menstruation, pregnancy and lactation. Marathon runners, ballet dancers and thin models can have as low as 7% BF. These women usually cease to menstruate.

The **normal** amount of body fat (expressed as a percentage of body mass) is between 25% and 30% in women and 18% and 24% in men. Women with over 32% body fat and men with over 25% body fat are considered obese.

Essential fat is the level below which physical and physiological health would be negatively affected.

In the Western world, obesity is probably the most prevalent disturbance related to food intake. Since the mid-1960s, there has been increasing concern that over-nutrition contributes to heart disease, cancer, diabetes and liver disease.

In many cases, persons with extra fat located around the middle ("apple" shaped) are at a higher risk for diseases, such as heart disease and diabetes, than those who carry weight around their hips and thighs ("pear" shaped). (A waist circumference of >88 cm for women or >102 cm for men defines abdominal obesity and substantially increased risk of obesity-related disease.)

For both men and women, a waist-to-hip ratio of 1.0 or higher is considered "at risk" for heart disease or other problems associated with being overweight (e.g. type II diabetes).

For men, a ratio of less than 0.90 is considered safe.

For women, a ratio of 0.80 or less is considered safe.

9 Digestive System Revision: Homework Exercise 6

1. List the organs of the digestive system and the accessory organs.
2. Name the three regions of the small intestine and describe what digestive and absorptive events happen in each region.
3. Describe the functions of each of the four tunics of the wall of the alimentary canal.
4. Draw up a table with three columns and ten lines (make up your own headings) that lists:

    (a) Each digestive enzyme
    (b) Which structure/cells secrete the enzyme
    (c) What the enzyme does

5. Name AND describe all the processes that constitute mechanical digestion.
6. Summarise the regulation of gastric secretion.
7. Summarise the events in the chemical digestion of carbohydrates.
8. Summarise the events in the chemical digestion of proteins.
9. Summarise the events in the chemical digestion of lipids.
10. Describe the basic anatomy of the liver (the functional unit and the cells).
11. State the functions of the liver (give details for three of those functions).
12. What is the effect of pepsin on protein?
13. Consider two men: one of 70 kg with 15% BF and the other of 100 kg and 12% BF.

    (a) Which one has a lower percentage of fat?
    (b) Calculate 12% of 100 kg and 15% of 70 kg. Which one has the greater mass of body fat?

14. In what clinical situations would skinfold measurements (or body mass index) provide useful information about the nutritional status of clients?

Lectures 9 and 10: Endocrine System & Hormones

1 Endocrine System Overview

Hypothalamus, pituitary (anterior & posterior parts), pineal, thyroid, 4 parathyroids, thymus, pancreas (islets of Langerhans, *not acini*), 2 adrenals, 2 gonads, placenta. (skin – vit D; gut – digestion controlling hormones, kidneys – erythropoietin, liver – angiotensinogen, adipose tissue – leptin)

Hormone: a molecule = a chemical messenger, made in very small quantities by endocrine cells/glands & released into circulation.
Some hormones (e.g. PTH, insulin, aldosterone) released in direct response to blood levels of ions/ nutrients.
Some (e.g. epinephrine, norepinephrine) released in response to neural signal.
Many released when directed to by a "releasing" or "stimulating" hormone from the hypothalamus.

2 Cell Communication

Cells can communicate with other cells by:

1. direct exchange of ions & small molecules through "gap junctions" between adjacent cells in contact.
2. paracrine communication (using cytokines, local hormones) through extracellular fluid between cells of a single tissue
3. synaptic communication – neurons release neurotransmitters at a synapse.
4. Endocrine communication uses hormones released into blood to coordinate cellular activities in distant tissues (=circulating hormones).

M. Caon, *Lecture Notes, Worksheets, and Exercises for Basic Anatomy and Physiology*, https://doi.org/10.1007/978-3-031-56296-9_7

- Chemical messengers may be BOTH hormones AND neurotransmitters (e.g. noradrenaline, dopamine).
- Some neurons secrete hormones (e.g. hypothalamus secretes *oxytocin* & *ADH* which are transported along the axon to posterior pituitary gland).

Endocrine glands/cells: secrete their hormones direct into blood stream. (≠ exocrine glands)

Target cell: point of action of a hormone – has a *specific* receptor site for that hormone.

Hormones persist in blood for from <1 minute to ~30 minutes. But because of "amplification" (see below) their *action* may persist for from ~20 minutes to several hours.

3 Chemical Structure of Hormones

Classified into 2 groups:

- amino acid (AA) based hormones
- steroid hormones.

- (see molecule structure diagrams)

1a. Amino acid derivatives. (AA are the structural units of proteins). "Catecholamines" (adrenaline, noradrenaline – from adrenal medulla, & dopamine – from hypothalamus) and thyroid hormone (T_4 & T_3) are all *derived from tyrosine.*

Melatonin is *derived from tryptophan.*

1b. Peptide hormones are chains of AA. Includes all h. released by hypothalamus, post. pit., heart, thymus, digestive tract, pancreas & most of ant. pit. h.

e.g. ADH & oxytocin (9AA long), glucagon (29AA), calcitonin (32AA), ANP (45AA), insulin (51AA), leptin (146AA), EPO (166AA), GH (191AA), PRL (198AA).

Glycoproteins e.g. TSH, LH, FSH from ant. pit. (>200 AA + carbohydrate side-chain)

Angiotensinogen (from liver) is 452 AA long

Amino acid-based hormones are transported dissolved in blood (except thyroid hormone which binds to a plasma protein).

2. Steroid hormones. Produced by

1. Gonads: testes (testosterone), ovaries (estradiol, progesterone),
2. Adrenal cortex (corticosteroids):

   Mineralocorticoids = aldosterone
   Glucocorticoids = cortisol, corticosterone
   Gonadocorticoids = testosterone, estrogens

3. Kidneys: calcitriol (active vitamin D_3) via skin & liver.

Steroid h. (lipids) are poorly soluble in blood so bind to plasma proteins for transport ⇒ stay in circulation longer. Liver eventually absorbs them, converts them to soluble form for excretion (in urine).

[a 3rd chemical group are the "local" hormones known as eicosanoids (lipids derived from arachidonic acid - a poly-unsaturated 20 C conditionally essential fatty acid). They include leukotrines, prostaglandins (which may become thromboxanes & prostacyclins) involved in paracrine communication]

4 Receptor Sites

To be affected by a hormone, a cell must have a *specific receptor* for that hormone. Receptors are either on outside of cell membrane (for AA based hormones) or inside (!) the cell.

Catecholamines (adrenalin, noradrenalin, dopamine) and peptide hormones are *NOT* lipid soluble (so are unable to penetrate cell membrane) so receptor proteins for these hormones are on outside of membrane.

Thyroid hormones cross membrane by a carrier mechanism & bind to receptors within nucleus or mitochondria.

Steroid hormones are lipid soluble so diffuse through cell membrane to bind to receptors in cytoplasm or nucleus and "switch on" a gene (i.e. prompts transcription to produce mRNA & thus proteins)

[Eicosanoids are lipid soluble so diffuse across membrane to receptor proteins on inside of cell membrane.]

5 Hormone action

Hormones can alter the physical shape of target proteins or biochemical properties of its target cells by:

1. (By activating appropriate genes in cell nucleus) Stimulating synthesis of an enzyme or a structural protein in the cytoplasm.
2. Activating or deactivating an enzyme by altering its shape or structure.
3. Altering plasma membrane permeability or electrical state by opening or closing membrane channels.
4. Stimulating mitosis.
5. Inducing secretory activity.

Proteins & peptides cannot penetrate cell membranes ➔ AA-based hormones (are 1st messengers) exert their effect via intracellular **second messengers** (e.g. cAMP, cGMP, Ca^{++}) after binding to a receptor on plasma membrane.

(i) hormone binds to a receptor which
(ii) activates a "G-protein" inside the cell which then moves along the membrane to
(iii) activate an enzyme that produces cAMP.
(iv) cAMP (the 2nd messenger) triggers a cascade of chemical reactions, producing increasing numbers of molecules at each step – amplification !

(theoretically receptor binding of a single hormone molecule could generate millions of final product molecules!)

6 Control of Endocrine Activity by the Hypothalamus

The hypothalamus is a region of the brain, below the thalamus, posterior to the optic chiasma and anterior to the mammillary body. it connects to the pituitary gland by a "stalk" (called the infundibulum).

Hypothalamus *controls* the endocrine system and *integrates the activities* of the nervous & endocrine systems (in the following three ways:)

1. Hypothalamus secretes regulatory hormones that:

   (a) control endocrine cells in the ant. pit. gland. (releasing hormones e.g.

   GnRH = gonadotropin releasing h.
   GHRH = growth hormone releasing h
   TRH = thyrotropin releasing h
   CRH = corticotropin releasing h
   & inhibitory hormones that stimulate the anterior pituitary gland to secrete hormones)

   (b) In turn, the seven hormones released by the ant. pit. control the activities of endocrine cells in the thyroid; cortex of adrenal glands; & reproductive organs.
   (c) these endocrine organs then release hormones that target different parts of the body.

2. Hypothalamus produces ADH & oxytocin which are transported to post. pit. stored and released into blood from post. pit.
3. Hypothalamus contains "autonomic centres" that exert neural control (via impulses through nerve fibres) over endocrine cells of the medullae of the adrenal glands. (When "sympathetic division" of the autonomic nervous system is activated, adrenal medullae release hormones into bloodstream).

[thus hypothalamus releases GnRH which stimulates ant. pit. to release FSH (& LH) which stimulates testes to release inhibin & ovaries to release inhibin & estrogens (LH stimulates testes to release androgens & ovaries to release progesterone & estrogens]

Pituitary Gland (=Hypophysis)

Connects to hypothalamus by the infundibulum

(a) *anterior pituitary* (glandular tissue which communicates with hypothalamus via capillary plexus and portal veins). =adenohypophysis Produces 6 hormones which are released when ant pit is stimulated by the arrival of releasing hormones from the hypothalamus:

1. growth hormone (GH = somatotropic h.);
2. prolactin (PRL);
   & four "tropic" hormones (Tropic = "turn on or cause release")
3. TSH = thyroid stimulating h.,
4. ACTH = adrenocorticotropic h.,
5. FSH = follicle stimulating h.,
6. LH = luteinising h.).
7. (MSH = melanocyte stimulating hormone, and β-endorphin & met-enkephalin)

(b) *posterior pituitary* (= neural tissue: axons from hypothalamus enter the post pit. ie it is part of brain)

It stores & releases oxytocin (OT) & antidiuretic h. (ADH) that are made in hypothalamus. OT & ADH transported to post. pit inside axons from the hypothalamus.

(infundibulum + posterior pituitary = neurohyopophysis)

7 Thyroid Gland

Composed of spherical sacs called "follicles". Thyroid hormones (as thyroglobulin) made by follicular cells & stored in follicles (3–4 month's supply!).

TRH causes release of TSH.

Under influence of TSH (from ant pit), thyroglobulin from follicle is endocytosed by follicular cells. T_3 and T_4 are produced & diffuse from follicular cells into blood stream where they attach to globulins and albumins (plasma proteins) for transport.

Thyroid gland produces thyroxine (tetra-iodothyronine) (T_4 with 4 I atoms) 90% & tri-iodothyronine (T_3 with 3 I atoms) 10% (T_4 is converted to T_3 in target tissues).

T_3 affects all body cells except brain, spleen, testes, uterus.

T_3 causes:

- Increase in rate of ATP formation (in mitochondrian).
- Increases metabolic rate of cell by stimulating synthesis of enzymes concerned with glucose metabolism (➔ heat generated).
- Accelerates production of the protein "Na-K ATPase" (Na-K pump)
- In children T_3 essential for normal development of skeletal, muscular & nervous systems.

Parafollicular ("C") cells of the thyroid produce **calcitonin** which aids in lowering of [Ca^{2+}] in blood (range 2.0–2.5 mmol/L). When plasma Ca^{2+} levels **rise**, more calcitonin produced which decreases Ca^{2+} levels by:

1. slowing release of Ca^{2+} from bone
2. stimulating Ca^{2+} excretion at kidneys

Acts in children & pregnant women.

8 Parathyroid Glands (Two Pairs of)

Secrete parathyroid hormone when plasma levels of Ca^{2+} **drop**. PTH increases plasma Ca^{2+} by:

1. accelerating release of Ca^{2+} from bone (by increasing osteoclast numbers)
2. reducing Ca^{2+} deposition in bone (inhibits osteoblasts)
3. enhancing reabsorption of Ca^{2+} from filtrate in kidneys
4. stimulating formation & release of calcitriol (=vit. D) in kidneys (which in turn enhances absorption of Ca^{2+} from gut)

9 Adrenal Glands (Two of)

Firmly attached to superior portion of each kidney. Project into peritoneal cavity. Divided into superficial adrenal **cortex** (3 zones) and deeper adrenal **medulla**.

(a) Adrenal cortex produces **cort**icosteroids from cholesterol (>24 of them!). Either "mineralo**cort**icoids" (e.g. aldosterone) or "gluco**cort**icoids" (e.g. cortisol, corticosterone) or "gonado**cort**icoids" (estrodiol & the weak androgens androstenedione, dehydroepiandrosterone).

Corticosteroids determine which genes are transcribed in the nucleus ➔ enzymes produced.
Glucocorticoids increase rate of glucose formation from fatty acids & proteins, & glycogen formation. Adipose tissue responds by releasing fatty acids into blood. (Topically applied glucocorticoid creams reduce inflammation)
Adrenal androgens (from zona reticularis) *in women* promote muscle mass, blood cell formation & libido.
Cortex (zona glomerulosa) produces **aldosterone**.
If plasma [Na^+] (137–143 mmol/L), blood volume or blood pressure drop, or [K^+] (3.2–4.3 mmol/L) increases, then aldosterone is secreted.
Aldosterone is also released when "angiotensin II" is present in plasma.
Aldosterone causes Na^+ to be reabsorbed in kidneys (& sweat glands, saliva & pancreas) in exchange for K^+.
Furthermore, if normal levels of ADH are present, water will be reabsorbed from kidney too (water "follows" Na^+ osmotically).
ADH also causes contraction of smooth muscle in blood vessels.

(b) Adrenal medulla produces catecholamines: adrenaline & noradrenaline (= epinephrine & norepinephrine).

Noradrenalin produces vasoconstriction
Adrenalin produces vasoconstriction in skin & viscera, but vasodilation in skeletal muscle.
Both cause bronchodilation, increase HR & contractility.

10 1% of Pancreas (Islets of Langerhans)

alpha (α) cells produce glucagon which targets the liver and causes:

- breakdown of glycogen to glucose (glycogenolysis)
- synthesis of glucose from lactic acid & from fats & amino acids (gluconeogenesis).
- Release by the liver cells of glucose into the blood

i.e. Glucagon ***raises*** blood sugar level.

beta (β) cells produce insulin which causes:

- enhanced membrane transport of glucose into body cells (esp. muscle & fat cells).
- Inhibits glycogenolysis.
- Inhibits gluconeogenesis.

(Once glucose is inside cells, insulin:

- Catalyses oxidation of glucose for ATP production
- Promotes glycogen formation in cells
- Converts glucose to fat)

i.e. Insulin ***lowers*** blood sugar level (range 6.1–8.3 mmol/L) (& promotes protein synthesis & fat storage).

[estimated that 300×10^6 people will have type II diabetes by 2025. Oral therapies are: increase insulin secretion (sulphonylureas); increase insulin sensitivity (thiazolidinediones); reducing hepatic glucose production (metformin); delaying absorption of glucose from GIT (α-glucosidase inhibitors)]

11 Gonads (etc)

(a) ovaries produce estrogens (e.g. estradiol), inhibin and progesterone.
(b) Testes produce androgens (e.g. testosterone) and inhibin.

Kidneys Erythropoietin is produced by interstitial fibroblasts in close association with the peritubular capillary and proximal convoluted tubule. EPO signals red bone marrow to increase production of rbc. Kidneys also produce calcitriol, and renin (an enzyme which converts angiotensinogen to angiotensin I).
Pineal gland produces melatonin (which has a mysterious function).
Atria of heart produce atrial natriuretic peptide (ANP) which blocks renin secretion.
(**Thymus** produces thymopoietins & thymosins which promote the maturation of T cells.
Leucocytes release leucotrienes, most tissues produce prostaglandins.
The intestines produce a variety of hormones that coordinate the activities of the digestive system.
Placenta produces estrogens, progesterone & hCG).

12 Homeostatic Role of Hypothalamus in Water Balance (Negative Feedback)

Stimulus: increase in plasma osmotic pressure due to

1. Large intake of salt, or
2. Dehydration.

Receptor: osmoreceptors in hypothalamus
Response: Hypothalamus stimulates post. pit. to release more ADH from axons in post. pit. & from there into blood stream.
Effector: collecting duct of nephron of kidney.

Collecting duct becomes permeable to water, more water is reabsorbed from "urine filtrate" (into extracellular fluid, then into plasma), small volume of concentrated urine produced. Also ADH stimulates the thirst centre ➔ we drink water.

If the plasma osmotic pressure decreases due to large intake of water (say), less ADH released, less water reabsorbed from filtrate, large volume of dilute urine produced.

Effect: plasma osmotic pressure is returned to within healthy range.

13 Additional Information

Diabetes

Diabetes is a chronic illness associated with both microvascular complications such as nephropathy, retinopathy, and neuropathy; and macrovascular complications including cardiovascular disease, stroke, and peripheral vascular disease. These complications are associated with reduced quality of life and increased morbidity and mortality and contribute significantly to the nations' healthcare costs. Life expectancy for a patient with Type 2 Diabetes Mellitus (T2DM) between the ages of 40–70 years is reduced by approximately 30–40%, yielding a loss of 8–10 years of life.

It is well recognized that the prevalence of T2DM has increased dramatically in recent decades, and this is generally attributed to environmental and behavioural factors such as diet and physical inactivity. The International Diabetes Federation estimates that there are approximately 248.8 million diabetics worldwide in 2010. A recent survey in China estimated that there were 92.4 million adult diabetics in China, and an additional 55.8 million with 'pre-diabetes', making China the second largest diabetic population in the world after India

Current pharmacologic treatments for T2DM include a diverse range of antidiabetic drugs.

However, a number of these are associated with clinically important and, in rare cases, life-threatening side effects, such as hypoglycemia (sulfonylurea and insulin), lactic acidosis (metformin), weight gain (sulfonylurea, insulin, and thiazolidinediones (TZDs), fluid retention and heart failure (insulin, TZDs), and gastrointestinal side effects (metformin, exenatide), potentially limiting their long-term use.

[Cure for T2DM = eat less and exercise more.]

Normal blood glucose = 6.1–8.3 mmol/L

Hypoglycaemia = blood glucose <3.3 mmol/L with symptoms present (or <2.7 mmol/L)

Hyperglycemia = BG >22 mmol/L

Insulin given sc, or orally (metformin), or by lung inhalation!

Hypoglycaemia = blood glucose <3.3 mmol/L with symptoms present (or <2.7 mmol/L)

Normal blood glucose = 6.1–8.3 mmol/L

Hyperglycemia = BG >22 mmol/L

Type 2 diabetes is present in ~6% of adult western population! (& incidence is growing at 6% per year)

Type 2 diabetes is most common form (85–90% of pt) of diabetes.

Type 2 diabetes is associated with obesity in 80% of afflicted pt. (most important contributor to pathogenesis of disease)

Major cause of death in pt with type 2 diabetes is macrovascular disease (coronary artery, cerebrovascular & peripheral vascular). In addition microvascular disease (retinopathy, nephropathy & neuropathy) has significant effect on quality of life.

Glycosylated Haemoglobin A_{1c} 5.6–7.5 mmol/L (Hb A_{1c})

Providing red cell lifespan is normal, Hb A1 measures mean blood glucose concentration over the preceding 60 days – i.e. half-life of red cell.

Some assays measure total glycosylated haemoglobin whilst others measure Hb A1c produced by glycosylation of the N-terminal valine of the B-chain of haemoglobin.

The relationship values of GHB and PG is defined in the table below:

Glycosylated haemoglobin (HBA(1c)) (%)	Approximate mean plasma glucose (mmol/l)	Approximate mean plasma glucose[a] (mg/dL)
4	3.5	65
5	5.5	100
6	7.5	135
7	9.5	170
8	11.5	205
9	13.5	240
10	15.5	275
11	17.5	310
12	19.5	345

[a]Mean blood glucose results are 10–15% lower. Most blood glucose meters are calibrated to read as plasma glucose

(males) Inhibin (produced by sustentacular cells of testes) released when sperm count is high. It inhibits release of GnRH (by hypothalamus) & release of FSH (by ant pit)

(females) inhibin released by granulose cells to inhibit FSH release

The reduction in the circulating levels of estrogen during menopause is associated with osteoporosis (loss of bone mass & deterioration of microarchitecture of bone tissue) ➔ increase in bone fragility and bone fracture. Treat with Ca & vit D & drugs (e.g. biphosphonate, calcitonin, selective serum estrogen modulator (SERM) & monitoring BMD.

14 Endocrine System Revision: Homework Exercise 7

1. (a) What are hormones?
   (b) Describe their chemical classification (and give an example of each)
2. Briefly describe the anatomical relationship between the hypothalamus and the pituitary gland.
3. What are the three ways that the hypothalamus controls the endocrine system and integrates the activities of the nervous and endocrine systems?
   (a)
   (b)
   (c)
4. Briefly describe the structure of the pituitary gland and name the two hormones stored in the posterior pituitary.
5. How are hormone receptors related to hormones?
6. How does the difference in hormone chemical structure determine where their receptors are located?
7. Describe an example that shows how the hypothalamus is able to control the secretion of a hormone from a "subservient" endocrine gland.
8. Copy and complete the table below: list the major one (or two) hormones produced; state a brief function for the hormone.

Gland	Hormone(s)	Function
Thyroid		
Parathyroid		
Adrenal cortex		
Adrenal medulla		
Pancreas		
Testes		
Ovary		
Pineal		
Anterior pituitary		
Posterior pituitary		
Thymus		

9. How do hormones exert their influence on target cells?
10. Describe how "second messengers" allow amino acid-based hormones to exert their influence on target cells.

Lectures 11 and 12: Urinary (Renal) System

Almost 12% of Australia's population has chronic kidney disease (CKD). In 2020-2021, haemodialysis treatment for end-stage kidney disease (ESKD) was the most common reason for hospitalisation in Australia, constituting 14% of all hospitalisations.

- Diabetes is the leading cause of CKD (and obesity → diabetes).
- A majority of patients with CKD do not survive to end-stage renal disease (ESKD needs dialysis or transplant).
- CKD and diabetes are associated with increased risk for mortality and cardiovascular (CV) disease.

1 Introduction

Two kidneys, two ureters, one bladder, one urethra (2–3 cm in females, ~20 cm in males and passes through the prostate gland)

Functions

1. Major excretory organs of the body (organic wastes, urea, uric acid, creatinine).
2. Regulate volume of blood (and hence blood pressure).
3. Regulate blood pH (loss of H^+ or HCO_3^-; deaminate amino acids to produce toxic NH^+_4 (excreted) and HCO_3^- retained in blood).
4. Maintain blood osmolarity at ~300 mosml/l (by Na^+, K^+, Cl^-, Ca^{++}, Mg^{++}, SO_4^{2-}, PO_4^{3-}) and healthy electrolyte concentrations.
5. Conserve nutrients by re-absorbing them from the filtrate (glucose, amino acids).
6. Produce the enzyme **renin** (which catalyses the formation of angiotensin I).

M. Caon, *Lecture Notes, Worksheets, and Exercises for Basic Anatomy and Physiology*, https://doi.org/10.1007/978-3-031-56296-9_8

7. The hormone **erythropoietin** (stimulates red blood cell (RBC) production in the marrow) is produced by interstitial fibroblasts in the kidney.
8. Produce **calcitriol** (active vitamin D_3).
9. Provide temporary storage for urine.

2 Kidney Anatomy

The kidneys are located between the body wall and the parietal peritoneum (**retroperitoneal**).

The **renal hilus** is the entry point for the renal artery, renal vein, lymphatics and nerves.

The fibrous **renal capsule** encases the kidney.

Internal Anatomy

(a) Cortex: superficial light-coloured tissue. All glomeruli are located here.
(b) Medulla: darker tissue containing ~8 "**pyramids**" – base towards cortex, apex (**papilla**) towards the pelvis. Pyramids have a striped appearance due to parallel urine-collecting tubules. Renal "columns" – with blood vessels – separate pyramids.
(c) Pelvis: a tube lined with smooth muscle, continuous with the ureter, which divides into two to three **major calyces**, each of which branches into several **minor calyces,** which form a cup around a papilla.

3 The Nephron

Nephron = blood-processing functional unit of the kidneys, > million per kidney

Eighty percent are **cortical nephrons** (located entirely – except for a small part of L. of H. – in the cortex).

Twenty percent are **juxtamedullary nephrons** (have long L. of H. extending deep into the medulla and have a vasa recta) – important in the production of concentrated urine.

Nephron = glomerulus + renal tubule.

1. Glomerulus is the vascular part of the nephron. The endothelium of glomerular capillaries is *fenestrated*, i.e. porous ⇒ large amounts of solute-rich, protein-free liquid pass from the blood into Bowman's capsule, thence to proximal convoluted tubule (PCT).
2. Tubule cells absorb ions from or secrete into the filtrate.

Renal tubule (3–5 cm long) = Bowman's capsule + proximal convoluted tubule + loop of Henle + distal convoluted tubule.

The distal convoluted tubule (DCT) of ~8 tubules empties into a common **collecting duct.**

3. The loop of Henle is "hairpin" shaped. The descending limb (thin segment) has a squamous epithelium, so it is *freely permeable to water*. The ascending limb may have some thin segment then becomes a thick segment (has cuboidal epithelial cells), which is *impermeable to water*.

4 Blood Supply

Descending aorta → renal artery → interlobar arteries (pass between pyramids) → arcuate arteries (arch over bases of pyramids) → cortical radiate artery (supply cortical tissue) → afferent arteriole (enters the glomerulus) → **glomerulus** (a capillary bed)

Glomerulus → efferent arteriole (small diameter, leaves glomerulus) → <u>either</u> peritubular capillaries (surround proximal & distal convoluted tubules) <u>or in juxtamedullary nephrons</u> **vasa recta** (a bed of permeable capillaries that surrounds Loop of Henle) → interlobular vein.

Nephrons have two capillary beds separated by efferent arteriole: the first bed (glomerulus) produces the filtrate, while the second (peri-tubular capillaries) carry away the water and molecules re-absorbed from the filtrate.

5 Bowman's Capsule and Glomerulus

(Renal corpuscle = glomerulus + Bowman's capsule)

Bowman's capsule surrounds the capillaries of the glomerulus. Gaps in the plasma membrane of the endothelial cells of glomerular capillaries make them <u>very</u> porous (100× compared to those of other capillaries).

Since an efferent arteriole has a smaller diameter than an afferent arteriole, blood in the glomerulus is under *high* hydrostatic pressure (55 mm Hg compared to 40–20 mm Hg in most capillaries).

Net filt press = G.H.P – G.COsP – B.C.Fl.P

Net filt press ≈ 55 – 30 – 15 = 10 mm Hg

Net filtration pressure (~10 mmHg) remains high throughout the length of the capillary – no venous end to the capillary.

High pressure forces liquid and solutes (wastes, ions, glucose, amino acids, fatty acids, vitamins) through the pores in fenestrated capillary walls, across the basement membrane (lamina densa), through the "filtration slits" of **podocytes** that surround a capillary and into the Bowman's capsule (proteins and blood cells remain in the blood).

(The Basement Membrane is an anionic barrier to albumin, in **diabetes** evidence that proteoglycan content of Glom BM is diminished so filtration barrier degraded.)

6 Autoregulation

Within limits, the kidney is able to maintain constant blood flow into glomerular capillaries by three mechanisms:

1. When BP increases, the afferent arteriole is stretched. It responds by constricting (a myogenic response), which decreases blood flow into the glomerulus, which causes the glomerular filtration rate (GFR) to decrease.

   When BP falls, the afferent arteriole can be dilated to increase GFR.

   Juxtaglomerular apparatus (= macula densa + granular cells)

   The JGA is where the distal end of the ascending limb of L. of H passes between (and touches) the afferent (and efferent) arterioles.
2. At this point, the cells of the *DC tubule* are called the **macula densa**. Macula densa cells are *chemoreceptors* that respond to changes in solute (Na^+ & Cl^- ions) concentration in the filtrate in the DCT. If [Na^+] and [Cl^-] are high:

   - GFR too high or
   - High blood pressure

   The macula densa sends a *paracrine* message (either ATP, or adenosine) to the afferent arteriole to constrict, which decreases blood flow to the glomerulus (which decreases glomerular pressure, which decreases GFR, which decreases ion concentration because more of them can be re-absorbed).

   If GFR is low ⇒ the macula densa sends a paracrine message (NO), which inhibits the action of adenosine triphosphate (ATP) and adenosine on the afferent arteriole. This increases blood flow and pressure in the glomerulus as well as GFR.
3. Adjacent to the macula densa, the afferent *arteriole walls* have specialised cells called **granular cells** (=juxtaglomerular cells) that produce **renin**. Granular cells are *mechanoreceptors* & sense blood pressure in afferent arteriole (respond to low pressure by releasing renin).

   Renin is an enzyme that converts angiotensinogen to angiotensin I; angiotensin-converting enzyme (ACE) in lung capillaries converts A1 to **angiotensin II**, which causes the constriction of systemic arterioles (which causes blood pressure to increase, which causes GFR to increase, which causes ion concentration in the filtrate to rise).

   Angiotensin II also (1) causes the adrenal cortex to release **aldosterone** and (2) stimulates thirst.

7 Urine Formation

(a) Filtration: in the glomerulus, blood cells and proteins are too big to pass through a filter. The Bowman's capsule receives everything else – water, nutrients, electrolytes and metabolic waste products – which is called the **filtrate**.
(b) Re-absorption from the filtrate (to reclaim useful substances for the blood) occurs in the renal tubules.
(c) Secretion from peritubular capillaries into the renal tubules occurs. Undesirable substances are disposed of this way, and blood pH is adjusted.
(d) The concentration of the filtrate may be increased in the collecting ducts. The walls of collecting ducts are impermeable to water, except when antidiuretic hormone (ADH) is present. **Antidiuretic hormone** (from the posterior pituitary) allows water to leave the filtrate through the wall of collecting ducts and causes filtrate volume to decrease and urine concentration to increase.

8 Filtration

Fenestrated capillaries of the glomerulus are more permeable than other capillary beds.

Glomerular filtration rate (GFR): ~120–125 ml/min. GFR $\propto$ net filtration pressure.

Approximately 20% of plasma passes into BC

- 180 litres of filtrate containing 25 mole Na^+ are produced daily.

Molecules smaller than 3 nm (water, glucose, amino acids, nitrogenous waste (urea, uric acid, creatinine), some albumin) and electrolytes pass through a filter.

Must produce 450 ml urine per day to eliminate soluble wastes.

9 Re-absorption

This returns most of the water and required solutes from the filtrate to the blood.

- Na^+ is actively re-absorbed.
- Water moves osmotically.
- Cl^- follows its electrical gradient.

PCT (surrounded by porous peritubular capillaries) re-absorbs:

- ~all glucose (secondary active transport)
- ~all amino acids (secondary active transport)
- 65% Na^+ (active transport)
- 65% H_2O (passive)

- 90% HCO_3^- (passive)
- 50% Cl^- (passive)
- 50% K^+
- 50% urea (passive by diffusion)
- ~100% uric acid
- 100% albumin (endocytosis)

If the parathyroid hormone (PTH) is present, less PO_4^{3-} is absorbed.
L. of H. re-absorbs:
From the *descending* limb: 15% H_2O
And from the *ascending* limb:

- 25% Na^+ (active)
- 25% K^+ (active, but it leaks back into the filtrate)
- 35% Cl^- (active)

DCT and collecting duct re-absorbs:

- Na^+ and Cl^- (active)
- ~15% H_2O if ADH is present
- More Na^+ from the end if aldosterone is present.
- Ca^{++} if parathyroid hormone is present.
- If ANP is present Na^+ re-absorption is inhibited

(**Diuretics** are substances that cause elevated urine flow because they slow renal re-absorption of water (causing **diuresis**). Most diuretics act by interfering with the re-absorption of Na^+.

(Caffeine inhibits Na^+ re-absorption; alcohol inhibits the secretion of ADH).

10 Secretion

Around 80% of plasma (and some unwanted substances) remains in the blood as it passes out of the glomerulus and moves into peritubular capillaries.

Some substances, rather than being filtered from the blood in the glomerulus, are **secreted** from the blood of peritubular caps into the filtrate:

- Ions: K^+ secreted into the DCT and collecting ducts under the influence of aldosterone (Na^+ is absorbed, and K^+ is secreted), Ca^{++} and PO_4^{3-}
- Certain drugs (penicillin, aspirin, phenobarbital, methotrexate, morphine, frusemide etc.), hormones and organic ions.
- Histamine and neurotransmitters
- Urea and uric acid in the DCT (after they have been re-absorbed passively in the PCT)
- H^+ in the PCT and DCT (in exchange for Na^+)
- Either H^+ (if blood pH falls) or HCO_3^- (if blood pH rises) throughout tubules

- NH^{+}_{4} in exchange for Na^{+} in the PCT and DCT
- Cl^{-} in exchange for HCO_3^{-}

11 Urine Concentration and Volume

The filtrate entering the descending limb of L. of H. is iso-osmotic (300 mosmol/L) with blood.

The osmolarity of the interstitial fluid of the medulla increases from ~400 mosmol/L in the outer medulla to ~1200 mosmol/L in the inner medulla.

As the descending limb is *permeable* to water (impermeable to solutes except for urea) and the interstitial fluid of the medulla is hyper-osmotic (due to urea), urea diffuses **in** and water diffuses **out** of the descending limb L. of H into the interstitial fluid (so the filtrate volume decreases). The osmolarity of the filtrate in the renal tubule increases to 1200 mosmol/L.

As the ascending limb is *impermeable* to water, Na^{+} is actively transported out of the filtrate into the interstitial fluid (Cl^{-} follows passively). The osmolarity of the filtrate decreases to 100 mosmol/L, volume being unaltered.

- Dilute urine: 15–29 ml/min of urine at a concentration of 65 mosmol/l urine enters the pelvis.

  If ADH is present (it usually is), the cells of the collecting duct become permeable to water.

- Concentrated urine: 1 ml/min up to 1200 mosmol/L

  Distal parts of collecting ducts are permeable to urea (and have membrane transporters for urea). Urea diffuses out into interstitial fluid (medulla), making it hyper-osmotic (from here, urea diffuses back into the descending limb L. of H. (other downstream parts of the tubule are impermeable to urea) and again travels with the filtrate into the collecting duct).

Vasa Recta

(1–2% of blood flows into the vasa recta)

The structure of the vasa recta (VR) allows the osmolarity of blood in the vessel to increase from 300 to 1200 mosmol/l (Wow! Blood becomes hyper-tonic). As **blood** from the efferent arteriole descends from the cortex to the medulla in the vasa recta, it loses some water and **gains** lots of Na^{+}, Cl^{-} and K^{+}. As blood ascends back to the interlobular veins, it gains lots of water re-absorbed from the filtrate in the descending limb of L. of H. (and loses some Na^{+}, Cl^{-}), so the osmolarity of blood in VR decreases from 1200 to 300 mosmol/L (becomes isotonic again).

This prevents the osmotic gradient in the interstitial fluid of the medulla from dissipating due to blood flow and water moving from the filtrate into the interstitial fluid. The water that leaves the descending limb of L. of H. enters the (ascending) vasa recta by osmosis.

12 Urine Storage and Elimination

The bladder is closed by the involuntarily controlled internal urethral sphincter. When relaxed, urine passes into the upper part of the urethra.

Voiding (**micturition**) occurs when the voluntarily controlled external urethral sphincter is relaxed.

Urinary incontinence – inability to prevent the leakage of urine.

Urinary retention – when the bladder is unable to expel its urine (catheter inserted in the urethra to drain the urine)

Polyuria = excess urine production (≫2 L/day)
Oliguria = inadequate urine production (50–500 ml)
Anuria = negligible urine (0–50 ml/day)
Uremia = the accumulation of nitrogenous waste (urea) in the blood due to end-stage renal disease that alters fluid, electrolyte and acid-base balance and produces symptoms (hypertension, anaemia, osteodystrophy).
Albuminuria = albumin in urine (bad) **Haematuria** = RBC in the urine

(Blue urine! due to ingesting methylene blue)

13 Urine Analysis

This is a common, economical and non-invasive diagnostic test to gather information about kidney/lower urinary tract disease. Drug testing! Normal results were used to exclude some diagnoses.

Urinalysis

Osmolality	50–1200 mosmol/kg
Specific gravity	1.003–1.029
Glucose	Negative
Bilirubin	Negative
Ketone	Negative
Occult blood	Negative
RBC	0–3/5 per HPF
pH	4.5–7.8
Protein	Negative
Urobilinogen	4–20 μmol/l
Nitrite	Negative
Leucocytes	0–5 per HPF
Creatinine	50–120 μmol/L

Twenty-Four-Hour Urine Sample

Ca	100–300 mg/day
Creatinine clearance	8.8–17.7 mmol/day
Renal calculus	No stones present
Na	40–220 mmol/day
Cl	110–250 mmol/day
K	25–125 mmol/day
HPO_4^{2-}	13–42.00 mmol/day
Mg	3–5 mmol/day
Protein	40–150 mg /24 hrs
Uric acid	1.48–4.43 mmol/day
Urea	450 mmol/day

(Test for human chorionic gonadotropin (hCG) in urine provides an early, reliable test for pregnancy. hCG appears in the blood soon after implantation.)

14 Additional Information

Creatinine is an acid waste of protein metabolism (increased in muscle atrophy/necrosis). Creatinine clearance is used to assess renal function (decreased in renal disease/failure).

[If creatinine in the blood >200 μmol/l ⇒ impaired renal function!]

During oliguria, the filtrate volume is so slow that renal transport mechanisms can re-absorb virtually all of Na and Cl, causing complete absorption of water as well. Consequently, urine production ceases and kidney failure follows.

In 2013, kidney-related disease directly or indirectly contributed to 11% of the deaths of Australians (more than deaths due to breast cancer, bowel cancer, prostate cancer, suicide and motor vehicle accidents combined). It is one of the few diseases in which mortality rates are worsening over time.

https://www.youtube.com/watch?v=SGGLq_vG1Gc
Young Doctors in Love (1982)

Episode "Diagnosis by Tasting Bodily Fluids"

Where madly crazed scientist Doctor Oliver Ludwig (Harry Dean Stanton) instructed a pathology class of young hospital interns, residents and interns about body fluids and orifices:

Osmotic Diuresis

Glucose is normally completely re-absorbed from the filtrate in the kidney tubules. However, if the level of glucose in the plasma is too high (uncontrolled diabetes), the kidneys cannot re-absorb all of it. Excess glucose contributes to the osmotic pressure of the filtrate; hence, less water than normal will be re-absorbed.

Diabetics with very high plasma glucose (>28 mmol/l) [normal fasting glucose level 3.9–6.1 mmol/l] produce abnormally high volumes of urine ≫2l (**polyuria**).

- Frequent urination and great thirst are symptoms of uncontrolled diabetes.

Diabetic nephropathy (an early sign of kidney disease) occurs when the kidneys leak a small amount of protein into urine (= microalbuminuria). This is due to thickening of glomerular basement membrane and partial depletion of anionic glycosaminoglycan moieties (mainly heparan sulphate) and resultant diminution of physiologic electrostatic charge barrier. The loss of glycoprotein anions, which present an ionic barrier to the albumin, allows the escape of albumin into glomerular filtrate (and later proteins) ➔ albuminuria/proteinuria. Albuminuria/proteinuria is toxic to tubular endothelial cells, leading to inflammation and scarring.

PCT is permeable to urea, so as the volume of the filtrate decreases due to the absorption of water, urea concentration in the filtrate increases. This promotes the passive re-absorption of urea along its concentration gradient. Rest of renal tubule is only slightly permeable to urea.

Blood in the afferent arteriole enters glomerular capillaries and is filtered into the Bowman's capsule.

Filtrate moves into the PCT.

Filtered blood leaves the glomerulus through the efferent arteriole and enters the capillary bed surrounding the PCT.

Material re-absorbed from the PCT ➔ interstitial fluid ➔ very porous capillaries around the PCT by rapid osmosis (colloid osmotic pressure and low (~13 mm Hg) hydrostatic blood pressure draw in water).

Filtrate moves into the loop of Henle.

Water re-absorbed from the descending limb ➔ interstitial fluid of the medulla ➔ into vasa recta capillaries by osmosis.

Electrolytes re-absorbed from the ascending limb ➔ interstitial fluid of the medulla ➔ into the descending limb of the loop of Henle to be transported deeper into the medulla.

Establishing a marked increase in the osmolarity of medullary interstitial fluid (1200–1400 mosml/l) due to:

1. The active transport of ions (Na^+, Cl^-, K^+) out of the thick portion of the loop of Henle (much of these ions re-enter the descending thin limb and the descending vasa recta to be transported deeper into the medulla, where most of them diffuse out into the interstitium again – those that do not continue up the ascending limb, from where they are actively transported out again!)
2. The active transport of ions (Na^+, Cl^-) from collecting ducts
3. The passive diffusion of large amounts of urea from collecting ducts

Vasa recta is a porous "counter-current" exchanger – descending limbs absorb sodium chloride and urea from the interstitium, and water flows out. (This makes the osmolarity of blood ~1200 mosmol/l!)

Ascending limbs lose almost all of the absorbed sodium chloride and urea to the interstitium while water diffuses back into the blood.

(In effect, the descending vasa recta absorbs what comes out of the ascending vasa recta, leaving the concentration gradient in the medullary interstitium unchanged.)

Making Concentrated Urine

When ADH is present, the collecting duct is permeable to water.

As filtrate moves through the duct, water moves by osmosis into highly concentrated interstitial fluid of the medulla.

(?Then what – this fluid moves into ascending vasa recta?)

15 Further Information

- Almost 5% of the US population have moderate to severely decreased kidney function, and ~20 million have kidney disease (proteinuria or GFR <90ml/min/1.73m^2)
- Diabetes is the leading cause of chronic kidney disease (CKD).
- The majority of patients with CKD do not survive to end-stage renal disease (ESRD).
- CKD and diabetes are associated with an increased risk for mortality and CV disease.
- Anaemia appears to be an independent risk factor for CV disease in this population.
- Hence, patients with CKD, diabetes and anaemia are at an increased risk of CV events!

Estimated GFR using the Modification of Diet in Renal Disease (MDRD) equation:

$$\begin{aligned} \text{eGFR} = {} & 186 \times (\text{serum creatinine})^{-1.154} \times (\text{age})^{-0.203} \\ & \times (0.742 \text{ if the subject is female}) \times (1.210 \text{ if the subject is black}) \end{aligned}$$

GFR (creatinine clearance) in ml/min estimated with the Cockroft-Gault formula:

$$GFR = \frac{(140\ \text{age}) \times (\text{actual body mass (kg)})}{72 \times \text{serum creatinine} (\text{mg} / \text{dL})} \times (0.85 \text{ for women})$$

For females, GFR = 0.85 × male GFR
GFR should be? >60 mL/min

Creatinine

0.07–1.00 mmol/L – the reference range for women
0.06–1.30 mmol/L – for men

Toxic Nitrogenous Waste
Ammonia – from the deamination of amino acids
Urea – from protein catabolism (3–8 mmol/L)
Uric acid – from the metabolism of nucleic acids
Creatinine – from creatine phosphate (0.06–0.13 mmol/L)

Blood urea nitrogen (BUN) is a blood test to measure amount of urea nitrogen in the blood. A raised BUN indicates kidney disease.

16 Urinary System Revision: Homework Exercise 8

1. Draw or trace a diagram of a kidney in the coronal section. Label the cortex, medulla, pyramids, columns, papilla, pelvis, major calyces and minor calyces.
2. Name the sections of the renal tubule in the order that filtrate passes through them.
3. What is meant by the terms glomerulus, vasa recta and Bowman's capsule?
4. With reference to the kidney tubule, distinguish between filtration and osmosis.
5. Which parts of the renal tubule are impermeable to water?
6. Name and briefly describe the four stages of urine formation.
7. Define and describe the function of the juxtaglomerular apparatus.
8. Describe the changes in the composition of the filtrate as it passes through the descending limb and then the ascending limb of the loop of Henle.
9. Explain how the kidney can produce concentrated urine when the body needs to.
10. Explain the effect of aldosterone on the composition of urine.
11. What is the role of ADH?
12. What is the role of aldosterone?
13. Describe the feedback response that results in:
    (a) ADH secretion
    (b) Aldosterone secretion
14. What is the function of each of the four structures of the nephron in urine formation?
15. Write a paragraph describing the blood supply to the kidney and nephron.
16. (a) How (and under what conditions) does the kidney excrete bicarbonate ions?
    (b) How (and under what conditions) does the kidney re-absorb or manufacture new bicarbonate ions?

Lectures 13–16: Nervous System

1 Overview

Cells = neurons (cell body + axon) and neuroglia (support cells)
Organs = brain, spinal cord, special sense organs, nerves

A. **Central nervous system (CNS)** = brain + spinal cord
It integrates, processes and co-ordinates sensory data and motor commands. The brain is the seat of higher functions (intelligence, memory, learning, emotion, personality).

B. **Peripheral nervous system (PNS)**
= neural tissue outside the CNS
It delivers sensory information to the CNS and carries motor commands from the CNS to the peripheral tissues.

Bundles of axons (or nerve fibres) = a nerve.
Cranial nerves connect to the brain.
Spinal nerves attach to the spinal cord.
A "nucleus" is a brain structure consisting of a relatively compact cluster of neuron cell bodies.
Ganglion = the concentration of nerve cell bodies outside the CNS (= a nucleus).
PNS is divided into two divisions:

B.1. Afferent (= sensory neurons, carries impulses towards the brain)
B.2. Efferent (= motor neurons, away from the brain)

The efferent division carries motor commands to the muscles and glands (effectors). It is subdivided into:

B.2(a) Somatic nervous system (SNS) and
B.2(b) Autonomic nervous system (ANS)

M. Caon, *Lecture Notes, Worksheets, and Exercises for Basic Anatomy and Physiology*, https://doi.org/10.1007/978-3-031-56296-9_9

The SNS controls the skeletal muscle (voluntary contractions and reflexes).

The ANS controls the smooth muscle, cardiac muscle, diaphragm and glands (involuntary, automatic function).

The ANS is sub-subdivided into:

B.2(b) (i) Sympathetic division (SD) (prepares the body for "fight or flight": for energetic muscular activity)

B.2(b) (ii) Parasympathetic division (PSD) (prepares the body for "rest and repose": conserves energy, promotes non-emergency function)

Senses

General senses: position in space, touch, pressure, temperature, pain

Special senses: located in special organs (vision, hearing, taste, smell, balance and acceleration)

2 Neurons

Nerve cells (neurons) conduct messages (= nerve impulses). They consist of:

- Dendrites: multiple processes that carry incoming impulses to the cell body as graded potentials.
- Cell body: containing nucleus and organelles.
- Axon: a single process that carries an **action potential** (= nerve impulse) away from the cell body.
- An axon is connected to the cell body by an "axon hillock", may be sheathed in "myelin" and ends with multiple terminal branches (may be 1 m long!).
- Telodendria: branching terminals of an axon (+ synaptic knobs) that synapse with another cell.

- Neurons are long-lived (100 years).
- They are amitotic (cannot replace themselves if destroyed).
- PNS nerve axons can regenerate if cut.
- CNS neurons do not regenerate.
- Neurons have a high metabolic rate (require abundant oxygen (O_2) and glucose – survive only a few minutes without oxygen).

Unipolar neuron: one attachment to the cell body, which is attached to a single combined dendrite-axon. Most sensory neurons, e.g. from the skin, are unipolar (afferent neurons) (cell bodies of somatic sensory neurons lie in dorsal root ganglia (outside the CNS).

Bipolar neuron: two processes from the cell body – one axon and one fused dendrite, in the retina and the olfactory receptor cell.

Multipolar neuron: many dendrites attached to the cell body, from which a single axon arises, which ends in synaptic terminals. This is the major type in the CNS.

Some are "interneurons" (association neurons) connecting to another neuron (located in the CNS).

All motor neurons are multipolar.

The cell bodies of *somatic* motor neurons lie **in** the CNS.
The cell bodies of some *autonomic* motor neurons are located in the sympathetic ganglia (outside the CNS).

Neuroglia/glial (Support) Cells – Six Types
In the CNS, there are four types (10× more abundant than neurons but comprise half of the mass of the brain):

- Ependymal cells (ciliated, circulate CSF)
- Astrocytes (control the chemical environment around neurons, hold neurons in place, function in synaptic transmission, form the blood-brain barrier (BBB))
- Microglia (monitor neuron health, are macrophages so can phagocytose microbes),
- Oligodendrocytes (produce myelin sheath and cholesterol)

In the PNS, there are two types:

- Satellite cells (surround cell bodies in the ganglia, supply nutrients, provide structure)
- Schwann cells (form myelin sheath) – form regeneration tubes to guide axon regrowth; phagocytise debris

The **myelin sheath** (a Schwann cell) surrounds (and insulates) axons – interrupted by "nodes of Ranvier" – and functions to prevent the leakage of charge from axons. Consequently, impulse propagation is much faster (120 m/s) (through "saltatory" conduction from node to node).

3 Neuron Function

Motor (efferent) neurons carry impulses from the brain to the muscles.

Sensory (afferent) neurons carry impulses from sense organs to the brain.

Interneurons (association neurons) lie in the CNS: they connect (synapse with) motor neurons to sensory neurons and make up approx 20–30% of all neurons.

Resting Membrane Potential
Cell membranes can maintain a difference in the distribution of ions between the interior of a cell and the extracellular fluid. This results in a difference in electrical charge. This charge difference is called the resting membrane potential.

- Inside cells, the concentration of large organic anions (proteins), phosphate (PO_4^{3-}), bicarbonate (HCO_3^-) and K^+ is high (Na^+ is low).

- Outside cells, the concentration of Cl^- and Na^+ is high (K^+ is low). Differences in ion concentrations result in the intracellular fluid of cells being negative with respect to extracellular fluid (i.e. the membrane is polarised).
- Differences in Na^+ and K^+ concentration are maintained by the cell membrane's sodium-potassium pump (this takes energy).
- Membrane potential most marked in the muscle and nerve cells (inside is −70 mV with respect to outside).
- Muscle and nerve cell membranes are excitable (i.e. produce action potentials in response to a stimulus that is above the threshold).

Graded Potential

A stimulus to a dendrite produces a "graded potential" (a small stimulus produces a small change in membrane potential while a larger stimulus a larger change), which travels a few millimetres in the membrane. That is, the resting membrane potential is changed to −55 or −60 mV.

A graded potential (if it is above a threshold) changes to an action potential at the "axon hillock".

Action Potential

Resting membrane potential (−70 mV) can "flip" to +30 mV (when a stimulus above the threshold is received) and then "flop" back to −70 mV. This flip-flop in membrane potential is called the **action potential**, which lasts ~3 ms and propagates along the length of the axon as a nerve impulse.

- Following an (above-threshold) stimulus to the cell membrane, Na^+ rushes into the cell through channels (flipping the potential from −70 mV to +30 mV). Immediately after this, K^+ rushes out of the cell (flopping the resting potential from +30 mV back to −70 mV).
- The voltage surge of depolarisation (flip-Na^+ in) and repolarisation (flop-K^+ out) of the membrane propagates along the axon and constitutes an electric current.

Synapse

It is the point of "contact" between two neurons or between a neuron and the muscle or gland cell it stimulates.

1. When an action potential arrives at the end of a transmitting axon (= the synaptic knob of a pre-synaptic neuron), extra-cellular Ca^{++} enters the knob (via voltage-gated channels). The knob has synaptic vesicles in the cytoplasm, which contain a neurotransmitter (a molecule, e.g. acetylcholine) that is released when triggered by Ca^{++}.
2. The gap between the transmitting cell's membrane and the receiving cell's membrane (= synaptic cleft) is crossed by the neurotransmitter (released from synaptic vesicles), which then lodges in the receptor of the ion channel.
3. The membrane of the receiving neuron/muscle cell/gland cell (= post-synaptic membrane) has gated Na^+ ion channels that open when a neurotransmitter molecule lodges in the receptor. This in-rush of Na^+ transmits the action potential to the target cell.

4. Then acetylcholinesterase hydrolyses ACh into acetate and choline; choline is absorbed by the neuron to synthesise more ACh.

Neurotransmitters (>100 types)
Most neurons produce two or more neurotransmitters for release at different times (or the same time), e.g. acetylcholine (ACh), norepinephrine (NE) (also dopamine, serotonin, gamma-aminobutyric acid (GABA), glutamate, adenosine triphosphate (ATP), glycine, nitric oxide (NO), carbon monoxide (CO), substance P etc.).

All somatic motor neurons release acetylcholine (ACh) at their synapses with skeletal muscle fibres. They are always excitatory. (Somatic motor pathways involve two motor neurons: the upper motor neuron (UMN), whose cell body is located in a CNS processing centre; the UMN synapses with a lower motor neuron, whose cell body lies in a brain stem nucleus or the spinal cord and whose axon ends in a skeletal muscle.)

4 Brain

The brain is composed of the cerebrum, diencephalon, brainstem and cerebellum [and cerebrospinal fluid (CSF)!])

http://ellenjmchenry.com/brain-hemisphere-hat/

The brain (and spinal cord) is surrounded by three **meninges** (membranes), which when inflamed ➔ meningitis:

1. *Dura mater*: a tough outer coat of fibrous connective tissue (adjacent to the skull, two layers separated by a thin layer of fluid except where sagittal and transverse sinuses – filled with venous blood returning to the internal jugular – pass between them). The dura extends into the brain as dural septa and holds the brain in place.
2. *Arachnoid mater*: attached to the inner surface of the dura mater. Web-like extensions tie the arachnoid to the pia mater (CSF fills the sub-arachnoid space between the arachnoid and the pia mater and cushions the brain against shock). Sub-arachnoid space also contains blood vessels.
3. *Pia mater* (contacts the brain): contains many tiny blood vessels.

Blood Supply
Right (R) and left (L) internal carotid arteries supply 80% of the cerebrum (via two middle cerebral arteries).

R & L vertebral arteries join to form the basilar artery (which branches into two posterior cerebral arteries). The "circle of Willis" is an arterial anastomosis encircling the pituitary gland and optic chiasma – it unites the brain's anterior and posterior blood supplies, connecting the basilar (vertebral) artery to the carotids; equalises blood pressure; and provides an alternate route for blood if occlusion occurs.

CVA (stroke) – block to blood circulation in the brain due to clot. The brain tissue dies. Survivors are typically paralysed down one side of the body and may have a sensory deficit (e.g. speech). Thirty-five per cent survive beyond 3 years.

Undamaged neurons may sprout new branches that grow into damaged areas and take over some lost functions.

Blood-Brain Barrier

The endothelial cells of the capillaries that serve the brain are seamlessly joined to each other by "tight junctions", i.e. no gaps between cells. A basement membrane surrounds the endothelial cells, and pericytes wrap around this. Also, astrocytes extend processes that envelop the capillaries and regulate their permeability. Hence, BBB capillaries are very selectively permeable.

BBB separates the brain from fluctuations in hormones, amino acids and ions that occur in the blood. Some hormones are neurotransmitters, and K^+ modifies threshold potential, so without BBB, brain neurons would "fire" uncontrollably.

O_2, carbon dioxide (CO_2), water (H_2O), fats, fatty acids and fat-soluble molecules (alcohol, nicotine, anaesthetics, steroid hormones, psychotropic drugs) can diffuse into brain interstitial fluid.

Glucose, essential amino acids and some electrolytes pass through passively by means of facilitated diffusion. Some viruses (measles) can enter the brain.

However, blood-borne metabolic wastes, toxins, proteins, small non-essential amino acids, K^+ and **most pharmaceuticals** are denied entry to the brain (non-essential amino acids and K^+ are actively pumped out of the brain).

BBB is absent in certain parts of the hypothalamus (to allow the release of and detection of hormones) and the vomiting centre of the brainstem. BBB is incomplete in newborn and premature infants.

The **choroid plexus** is part of BBB but contains capillaries that are porous. Instead, the ependymal cells that surround the capillaries are joined by continuous tight junctions. Plasma that filters through the choroid plexus forms CSF.

Cerebrospinal Fluid (CSF)

About 150 ml of CSF is found in (two lateral ventricles + the third and fourth ventricles and the central canal) and around the brain and the spinal cord. CSF has less protein, Ca^{2+} and K^+ and more Na^+, Cl^- and H^+ than plasma.

CSF solution exchanges freely with the interstitial fluid of the brain.

The brain is supported by (floats in) CSF. CSF cushions the brain and transports dissolved gases, nutrients, wastes and chemical messengers.

CSF circulates through ventricles (aided by ciliated ependymal cells in the ventricles), the central canal of the spinal cord and the sub-arachnoid space. CSF re-enters blood in the superior sagittal sinus via "arachnoid granulations". It is replaced every 8 hours.

Cerebrum

The cerebrum surface consists of ridges (gyri) and grooves (sulci) organised into lobes.

Sulci and gyri increase the brain surface area so more neurons can be accommodated. Humans have a great capacity for "higher functions".

It is divided into left and right cerebral hemispheres by *longitudinal fissure*.

The *central sulcus* separates the frontal lobe (with the pre-central gyrus) from the parietal lobes (with post-central gyri).

Each hemisphere has a superficial layer of "grey matter" called the **cerebral cortex**, lying over "white matter". Within white matter are "islands" of grey matter called **basal nuclei**.

Grey matter is predominantly neuron cell bodies.

White matter is myelinated axons (fibres).

1. Cerebral cortex: this is our *conscious mind* (it allows us to be aware of ourselves and our sensations and to communicate, remember, understand and initiate voluntary movements). Sulci divide each hemisphere into five sections called lobes:

   (i) Frontal lobe (higher intellectual functions) and four motor areas:

   - Primary motor area (= pre-central gyrus, controls contractions of the skeletal muscles)
   - Pre-motor cortex (co-ordinates learned movements, relays instructions to the primary motor area)
   - Speech (= Broca's area, regulates vocalisation)
   - Frontal eye field (part of the pre-motor cortex)

   (ii) Parietal lobes (sensory cortex = post-central gyrus)
   (iii) Temporal lobes (auditory cortex)
   (iv) Occipital lobe (visual areas)
   (v) Insula (equilibrium, gustatory)

   We can identify the following *functional* areas:

   - **Motor areas** (four) – all in the frontal lobe.
   - **Sensory areas** (eight) – in the parietal, temporal, insula and occipital lobes.
   - **Association areas** (eight+) in all lobes – these communicate (i.e. "associate") with motor, sensory and other multi-modal association areas and use past experience to analyse, interpret and act on sensory input.

- **Post-central gyrus** (of the parietal lobe) houses the primary somatosensory area. Neurons in this area receive information from general sensory receptors in the skin and skeletal muscle (and taste).
- **Pre-central gyrus** (of the frontal lobe) houses the primary motor cortex. Neurons (pyramidal cells) in these gyri allow us to consciously control skilled voluntary muscle movements.

- The cerebrum encloses the left and right lateral "ventricles" (= "spaces" filled with CSF).

2. Cerebral white matter (communication tracts of three types):

   (i) *Commissural fibres* connect corresponding areas in the two hemispheres (corpus callosum).

(ii) *Association fibres* connect different parts of the same hemisphere.
(iii) *Projection fibres* enter the hemispheres from the lower brain or cord areas.

3. Basal nuclei/ganglia are grey matter adjacent to the lateral ventricles.
   The main components are the striatum, both the dorsal striatum (= caudate nucleus + putamen) and the ventral striatum (nucleus accumbens + olfactory tubercle); globus pallidus; ventral pallidum; substantia nigra; and sub-thalamic nucleus.

(The basal nuclei do not initiate movement but regulate the initiation and termination of skeletal movements. Have a complex role in controlling movement by inhibiting the skeletal muscle tone, suppressing unwanted movements and co-ordinating slow sustained contractions of posture, e.g. controlling the subconscious swing of the arms and legs during walking and "picking up a pencil").

Diencephalon

It consists of the thalamus, hypothalamus, pituitary gland and pineal gland.

Links the (inferior) brainstem with the surrounding cerebral hemispheres (relays and processes info).

The thalamus surrounds the third CSF ventricle. It contains about 13 "nuclei" that receive sensory information before passing it onto basal nuclei and the cerebral cortex (our consciousness) – but not all of it! (It directs attention to the stimuli of interest and elicits emotional responses.)

The hypothalamus (with ~12 nuclei) lies below the thalamus and forms the floor of the third ventricle. It is found behind the optic chiasma and in front of mammillary bodies. It receives sensory impulses from somatic and visceral sensors and has receptors for osmotic concentration, blood glucose and temperature.

(The optic chiasma is the crossover of optic nerves II. Mammillary bodies relay olfactory information and contain reflex centres for eating and swallowing movements.)

The hypothalamus *is the main visceral control centre of body homeostasis*. It is the autonomic control centre from which orders flow to lower CNS centres for execution.

It has many homeostatic roles:

1. Heart activity and blood pressure
2. Emotional response centre (facial expression)
3. Body temp regulation
4. Regulation of food intake and gut movement
5. Regulation of urine output and thirst
6. Regulation of sleep-wake cycles
7. Production of two posterior pituitary hormones
8. Release of regulatory hormones to the anterior pituitary for secretion
9. Control of uterine contractions and milk ejection

The pituitary gland "dangles" inferior to the hypothalamus by the infundibulum (stalk).

The pineal gland produces the hormone melatonin, which regulates circadian rhythm and reproductive functions. It is not isolated by BBB!

Brainstem

It consists of the midbrain, pons and medulla oblongata. Centres in the brainstem produce rigidly programmed, automatic behaviours necessary for survival, e.g. respiratory centre (breathing rhythm), cardiovascular centre (force and rate of heart contraction, vasoconstriction), vomiting, coughing, hiccoughing and swallowing.

Cranial nerves (12 pairs associated with the brain)

I = olfactory nerves
II = optical nerves

The nuclei of the brainstem are associated with ten of the 12 cranial nerves.

Midbrain (corpora quadrigemina = superior colliculi, which receive and process visual info, + inferior colliculi, which receive and process auditory info)

The substantia nigra releases dopamine.

III = oculomotor nerves
IV = trochlear nerves

Pons contains many fibres running to and from the cerebellum.

V = trigeminal nerves (trigeminal neuralgia)
VI = abducens nerves
VII = facial nerves

Medulla oblongata (MO) – fibre tracts of the spinal cord between higher and lower neural centres pass through the medulla.

The point of "decussation of the pyramids" is in the medulla (90% of corticospinal pathways cross over in the MO; hence, each cerebral hemisphere controls the voluntary movements of the opposite side of the body).

The medulla is an autonomic reflex centre containing visceral motor nuclei (cardiovascular centre, respiratory centres, vasomotor centres).

VIII = vestibulocochlear nerves
IX = glossopharyngeal nerves
X = vagus nerves
XI = accessory nerves
XII = hypoglossal nerves
X = **vagus nerves**

The vagus nerve is the only cranial nerve that extends beyond the head and neck region.

Sensory function (taste, baroreceptors in carotid sinus, chemoreceptors in carotid & aortic bodies). Parasympathetic motor function (heart, lungs, digestive system activity, liver & kidneys)

Cerebellum

The cerebellum processes inputs from the cerebral motor cortex, various brainstem nuclei and sensory receptors. The cerebellum subconsciously provides precise timing and appropriate patterns of *learned* skilled skeletal muscle contraction for smooth co-ordinated movements, posture and agility (= "automatic pilot").

Cerebellar injury results in the loss of muscle tone and clumsy unsure movements.

[Cerebrum	(telencephalon)
Diencephalon	(between the cerebrum and the brain stem)
Midbrain	(mesencephalon – brainstem)
Pons	(metencephalon – brainstem)
Medulla oblongata	(myelencephalon – brainstem)
Cerebellum	(metencephalon)
Forebrain = cerebrum + diencephalon]	

5 Brain Function

The "**limbic system**" is a network of neurons that span large distances in the brain but work together. They are a functional grouping rather than an anatomical one.

It is our motivational system – the limbic system makes us want to do tasks.

Limbic System

It sits on the border between the cerebrum and the diencephalon. It includes the cingulate gyrus (of the frontal lobe), the dentate gyrus and the parahippocampal gyrus (of the temporal lobe), several other nuclei (amygdala, septal nuclei, two in the thalamus), the mammillary bodies, the olfactory bulbs and the fornix.

It is our emotional or affective brain. It:

- Recognises angry or fearful facial expressions
- Expresses our emotions through gestures etc.
- Triggers emotional reactions to smells
- Produces emotion-induced illness (high blood pressure, heartburn, psychosomatic illnesses)
- Allows emotion to override logic
- Allows reason to prevent emotional response
- Is involved in memory and learning

(Fibres from parietal and temporal lobes go to association areas in the frontal lobe that are part of the limbic system. Through these connections, sensory information can be invested with emotional and motivational significance! An animal's behaviour can be changed from savage to docile and back again simply by stimulating different areas of the limbic system (e.g. amygdala).

Reticular formation (reticular activating system)

- net-like interconnection of neurons throughout the brain stem that connects to the cerebrum, hypothalamus, thalamus, cerebellum and spinal cord
- receives and integrates all incoming sensory input(!)
- keeps the cortex alert and us aroused and conscious
- filters out repetitive, familiar or weak sensory signals but allows significant, unusual or strong stimuli to reach consciousness (like your mobile phone ring tone!)

6 Spinal Cord (SC)

- The spinal cord extends from the foramen magnum of the skull to L1/L2(below L2, the dorsal and anterior roots of the remaining spinal nerves continue within the vertebral column until they exit at the lumbar, sacral and coccygeal regions).
- Spinal reflexes initiate and complete at the spinal cord level.
- The spinal cord contains "ascending" nerve tracts (transmitting sensory information) and "descending" tracts (transmitting motor information).
- It is protected by bone, meninges and CSF.
- Thirty-one pairs of spinal nerves attach to the SC.

The SC is enclosed by a single-layer dura mater (spinal dural sheath). External to the dura mater is the epidural space containing fats and veins within the vertebral foramen. The SC ends at L1, but dural and arachnoid membranes (and the roots of some spinal nerves) continue inferiorly to S2. The sub-arachnoid space contains CSF and delicate spinal nerves (cauda equina) ➔ lumbar tap here.

SC Cross-Section

SC is divided by the anterior median fissure and the posterior median sulcus into left and right.

The **grey matter** of the SC has an "H" or "butterfly" shape; that is, the left and right halves are joined by "grey commissure".

The posterior/dorsal horn contains interneurons.

The anterior/ventral horns contain cell bodies of motor neurons.

The **white matter** (myelinated and unmyelinated nerve fibres) is divided into anterior, lateral and posterior "columns" (funiculi). Each funiculus contains several "tracts" made up of axons with similar destinations that connect the brain to the peripheral body (i.e. pathways). Most motor pathways cross from one side of the CNS to the other (decussate) in the brainstem.

Ascending pathways (spino-thalamic and spino-cerebellar tracts) conduct sensory information from the body to the brain (upward(!) and *afferent*).

Descending pathways (cortico-spinal and all tracts ending with "-spinal tracts") conduct *efferent* impulses from brain to body.

SC breaks above L1 ⇒ paraplegia.
SC breaks above ~C7 ⇒ quadriplegia.
SC breaks above C5 ⇒ respiratory failure and death?
CVA ⇒ hemiplegia?

Spinal Nerves

- 31 pairs issue from the SC (1 pr. for each vertebra)
- eight pairs from C1 to C7 (C4–C7 serve the arms)
- 12 pairs from T1 to T12
- five pairs from L1 to L5 (serve the legs)
- five pairs from S1 to S5
- one pair from the coccyx

Each spinal nerve is formed by a dorsal root and a ventral root (surrounded by dura mater) coming together in the vertebral foramen.

The <u>dorsal root</u> links to the posterior grey horn. It contains sensory fibres carrying incoming/afferent impulses from peripheral sensory receptors. These sensory fibres arise from neurons whose cell bodies are located in **dorsal root ganglia**.

The <u>ventral root</u> (with motor neurons carrying outgoing/efferent impulses) links to the anterior grey horn (autonomic NS efferents are also in the ventral roots).

Each spinal nerve is short (1–2 cm), then it divides as it **leaves** the vertebral foramen into dorsal ramus, ventral ramus, grey ramus, white ramus etc.

Note:

The **phrenic nerve** arises from spinal nerves C3, C4 and C5 and innervates the diaphragm.

The **sciatic nerve** (L4, L5, S1–S3) may be damaged by the improper administration of injection into the buttocks.

(Inborn) Spinal Reflexes

This refers to a rapid predictable motor response (unlearned and involuntary) to a stimulus (without processing by the brain).

1. Receptor
2. Sensory neuron (transmits afferent impulses to the CNS)
3. Integration centre in the CNS (spinal cord)
4. Motor neuron (conducts efferent impulses to an effector organ)
5. Effector (muscle fibre or gland cell)

Reflexes are used by physicians to assess the condition of the nervous system. Exaggerated, distorted or absent reflexes indicate the degeneration/pathology of specific nervous system regions.

Stretch Reflex

The stretch reflex (e.g. patellar reflex) automatically controls the length of the skeletal muscle.

"Muscle spindles" are proprioceptors within a muscle. They are special muscle fibres wrapped in sensory nerve endings. When the spindles in a muscle are stretched, afferent fibres carry impulses to the SC, where 2 synapses occur – one with a motor neuron and another with an interneuron; the motor neuron carries an impulse to the stretched muscle, causing it to contract (i.e. oppose the initial stretch), and the interneuron synapses with another motor neuron, which signals the antagonistic muscle to relax (i.e. to not oppose the contraction of the first muscle).In this way, muscle tone is maintained, and we can control our leg muscles and maintain standing posture (the patellar – "knee-jerk" – is a manifestation of a monosynaptic stretch reflex).

(The <u>tendon reflex</u> responds to increased stretch in the **tendon** organ. The sensory neuron synapses with two interneurons, one of which synapses with a motor neuron, which then causes muscular relaxation to relieve tendon stretch, and the other causes contraction in the antagonistic muscle.)

Cranial reflexes involve the cranial nerves. They are used to check for damage to the cranial nerves or their associated processing centres in the brainstem.

7 The Autonomic Nervous System

- The ANS stimulates the cardiac muscle, smooth muscle, diaphragm and glands. Hence, it controls heart rate (HR), vasodilation, blood pressure, body temperature, respiration, digestive function, urinary function, reproductive function and endocrine function.
- The ANS has two neurons in a motor pathway:
  - <u>neuron #1</u> (the pre-ganglionic neuron) runs from the CNS to a ganglion (then synapses), and
  - <u>neuron #2</u> (unmyelinated post-ganglionic neuron) runs from the ganglion to the effector (then synapses). The ganglion contains the cell body of neuron #2.
- The ANS has two divisions:
  - <u>"Sympathetic" division</u> (which mobilises the body during extreme situations, e.g. exercise, excitement, emergency) and
  - <u>"Parasympathetic" division</u> (maintains body activities, e.g. digestion, defecation and diuresis, and conserves energy).

(Some organs, like the eye, heart, pancreas, lungs, gut, bladder and reproductive system, are innervated by **both** SD and PSD.)

Sympathetic Division (SD)

SD fibres emerge from the thoracic and lumbar vertebrae (hence are "thoraco-lumbar" T1-L2).

SD has short pre-ganglionic fibres and long post-ganglionic fibres.

SD have ganglia that lie close to the SC.

SD supplies sweat glands, arrector pili muscles, and smooth muscle of blood vessels (vasoconstriction).

SD stimulates the adrenal gland to release NE and epinephrine (adrenaline) and the kidneys to release renin (an enzyme) into the blood.

Parasympathetic Division (PSD) = Vagus

PSD fibres emerge from the brain and the sacral spinal cord (hence are "cranio-sacral").

PSD has long pre-ganglionic fibres and short post-ganglionic fibres.

Most PSD ganglia lie in a visceral effector organ.

Ninety per cent of PSD impulses are via the two vagus nerves (X).

Parasympathetic Neurotransmitters and Receptors

All parasympathetic neurons release ACh as a neurotransmitter.

ACh is released by all ANS (both PSD and SD) pre-ganglionic fibres, i.e. at the first synapse (called cholinergic), to nicotinic acetylcholine receptors.

ACh is released by all ANS-PSD post-ganglionic fibres, i.e. at the second synapse (called cholinergic), to nicotinic or muscarinic receptors.

Parasympathetic effect on:

HEART – decreases HR
LUNGS – constricts bronchioles
EYE – constricts pupil
GUT – increases motility and secretions
REPRO SYS – causes sexual arousal

Sympathetic (stirs you up) effect on:

HEART – increases HR and strength of contraction
LUNGS – dilates bronchioles, increases respiratory rate and depth of breathing
EYE – dilates the pupil
GUT – decreases motility and secretion
URINARY – decreases blood flow to the kidneys
BLOOD VESSELS – assists in constriction (NE) or dilation (ACh)
SWEAT GLANDS – stimulates sweating (ACh)
SKEL MUSC – increases glycolysis and muscle tone
ADIPOSE TISSUE – assists in lipolysis

Two Neurotransmitters and Seven Receptors

Parasympathetic Neurotransmitters and Receptors

All parasympathetic neurons release ACh as a neurotransmitter.

ACh is released by all ANS (both PSD & SD) **pre-**ganglionic fibres, i.e. at the first synapse (called *cholinergic*), to nicotinic acetylcholine receptors.
ACh is released by all ANS-PSD **post**-ganglionic fibres, i.e. at the second synapse (called *cholinergic*), to nicotinic or muscarinic receptors.

[NE is released by most ANS-SD **post**-ganglionic fibres, i.e. at the second synapse (called *adrenergic*)
(some release ACh to muscarinic receptors).]
It is the type of *receptor* (not the neurotransmitter) that determines the response of the post-synaptic cell.

Cholinergic (Nicotinic and Muscarinic) Receptors
ACh binds (on the post-synaptic membrane of ganglionic neurons or on effector cells) to two types of (cholinergic) receptors: **nicotinic** and **muscarinic (M_1, M_2, M_3, M_4)**.
Nicotinic receptors bind ACh and are *always* stimulatory.
Nicotinic receptors are found on:

- All ganglionic neurons
- The hormone-producing cells of the adrenal medulla
- (And motor end plates of skeletal muscle cells)

Muscarinic receptors may be stimulatory *or* inhibitory.
Muscarinic receptors are found on all effector cells stimulated by post-ganglionic cholinergic fibres (= all parasympathetic target organs, a few sympathetic targets).

Sympathetic Neurotransmitters and Receptors

ACh is released by all ANS **pre-**ganglionic fibres, i.e. at the first synapse (called *cholinergic*), to nicotinic receptors.
NE is released by most ANS-SD **post**-ganglionic fibres, i.e. at the second synapse (called *adrenergic*) (some release ACh to muscarinic receptors).

Adrenergic (Alpha 1,2 and Beta 1,2,3) Receptors
Norepinephrine (and epinephrine) binds to five types of (adrenergic) receptors: α_1, α_2, β_1, β_2 and β_3. Binding may produce stimulation or inhibition depending on the target organ.

NE binding α_1 ➔ release of Ca and contraction of the smooth muscle in the blood vessels and sphincters in the gut
NE binding α_2 ➔ lowers the cAMP level in the cytoplasm, which inhibits
NE binding β_1 ➔ stimulates metabolic activity in the skeletal muscle, heart rate and force of contraction increases
NE binding β_2 ➔ inhibits smooth muscles, respiratory passageways dilate
NE binding β_3 ➔ lipolysis in the adipose tissue, free fatty acids (FFAs) into circulation

8 Additional Information

Knowing the location of receptor types allows the use of drugs to block nerve transmission across those synapses!

For example:

Atropine (an anticholinergic) binds to ACh muscarinic but not ACh nicotinic receptors to block PSD stimulation (but not SD stimulation).
Ephedrine (a sympatho-mimetic) binds to α adrenergic receptors to stimulate "feeling good".
A *beta-blocker* (an anti-hypertensive) binds to cardiac β_1 receptors to inhibit an increase in HR or the strength of contraction.
Ventolin (a bronchodilator) binds to lung β_2 receptors to mimic stimulation by NE.
LAS-3427 in development (a muscarinic antagonist bronchodilator) binds to lung M_3 receptors to block the normal response to ACh.

(The M1 cholinergic receptor is common in exocrine glands bound to G proteins, which use intracellular calcium as a signalling pathway.

M2 muscarinic receptors are located in the heart, where they act to slow the heart rate down to normal sinus rhythm after stimulatory actions of the sympathetic nervous system by slowing the speed of depolarisation. They also reduce contractile forces of the atrial cardiac muscle and reduce the conduction velocity of the atrioventricular node (AV node).

M3 muscarinic receptors are located at many places in the body, e.g. smooth muscles, endocrine glands, exocrine glands, the lungs, the pancreas and the brain. These receptors are highly expressed on pancreatic beta cells and are critical regulators of glucose homoeostasis by modulating insulin secretion. In general, they cause smooth muscle contraction and increased glandular secretions.)

Multiple Sclerosis is an autoimmune demyelinating disease of CNS

Current USA Statistics for Stroke Survival Rates

- Ten per cent of stroke victims recover almost completely.
- Twenty-five per cent of stroke victims recover with minor impairments.
- Forty per cent of stroke victims experience moderate to severe impairments requiring special care.
- Ten per cent of stroke victims require care in a nursing home or other long-term care facility.
- Fifty per cent die shortly after the stroke.
- Meanwhile, 7.6% of ischemic strokes and 37.5% of haemorrhagic strokes result in death within 30 days.
- While sub-arachnoid haemorrhage (SAH) represents only about 7% of all strokes, it is the deadliest – with more than 50% fatality rate. Of the survivors, approximately half will suffer permanent disability.

- Twenty-two per cent of men and 25% of women die within a year of their first stroke.
- Fourteen per cent of people who had a stroke or transient ischemic attack (TIA) will have another within a year.
- About 25% of stroke victims will have another within 5 years.
- Each year, 28% of people who suffer a stroke are under the age of 65.

Reflexes

Information about the environment around us (both internal and external) is detected through a variety of sensory receptors which convert (transduce) various stimuli into nerve action potentials. These action potentials are relayed to our CNS for processing.

Our bodies have receptors sensitive to mechanical, thermal, electromagnetic and chemical stimuli. Without any prior learning, relatively simple, inbuilt patterns of motor responses called REFLEXES can be exhibited. These reflexes are mediated over functional units called REFLEX ARCS.These responses to stimuli do not require voluntary intervention – that is, they do not require conscious efforts on our part to take place. They are, in the narrowest sense, "built in". For instance, there is a reflexive pulling away of one's hand from a hot object or the shutting of one's eyes as an object rapidly approaches one's face. In these cases, we may only be aware of the stimulus and the final event in the sequence (the reflex response). No matter how basic they appear to be, most reflexes are subject to alternation by learning. For instance, you can – if you want to – resist the impulse to remove your hand from a hot object or keep your eyes open when an object is hurtling rapidly towards your face. The testing of reflexes is an important and objective part of the examination of neurological functioning. Abnormal reflex responses can indicate specific pathologies in the CNS. Somatic reflexes involve skeletal muscles, whereas autonomic reflexes involve smooth muscles and/or glands.

The essential components of all reflex arcs (although most are more complex than this) are:

1. A SENSORY RECEPTOR, which reacts to a stimulus and transduces it into an electrical impulse
2. An AFFERENT (or SENSORY) neuron, which conducts transduced information from the receptor to the CNS
3. A CENTRAL region in the CNS where an incoming sensory impulse generates an outgoing motor impulse (here the impulse may be inhibited, transmitted or rerouted)
4. An EFFERENT (MOTOR) neuron, which conducts the integrated neural output from the CNS to an effector organ
5. An EFFECTOR ORGAN – the muscle or gland that responds to the stimulus.

Some reflexes have pathways where signals can pass via only two neurons, linked by one synapse. These are called MONOSYNAPTIC reflex arcs. Others are more complex, having one or more intermediary (internuncial) neurons in the reflex arc pathway. These are POLYSYNAPTIC reflex arcs.

Since delay (or inhibition) of the reflex may occur at the synapses, the larger the number of synapses in a pathway, the greater the time required to effect the reflex. (In humans and other mammals, the fastest impulses in the largest nerve fibres travel at a speed of over 120 m/s – i.e. over 400 km/h.)Many reflexes occur without the involvement of higher brain centres and will function as long as the spinal cord remains intact. Others require the involvement of functional brain tissues because many different inputs have to be evaluated before the appropriate reflex is determined. In some of the "spinal" "reflexes", while the brain is not required to produce a reflex response, it is frequently "advised" of spinal cord reflex activity and may alter it by facilitating or inhibiting the reflex.

Based on the distribution of effector organs in the body, reflexes are commonly recognised as belonging to one of two large groups:

1. **Somatic reflexes**: these are associated with the stimulation of the skeletal muscle by the somatic division of the nervous system. Some somatic reflexes require only spinal cord activity, e.g. the PATELLAR (KNEE JERK) and ACHILLES reflexes, while some require brain involvement as well, e.g. CORNEAL and PHARYNGEAL (GAG) reflexes.
2. **Autonomic (visceral) reflexes**: these are associated with visceral structures. These are mediated through the autonomic nervous system and are not subject to direct conscious control. They involve the regulation of such body functions as digestion, peristalsis, elimination, blood pressure and salivation.

9 Nervous System Revision: Homework Exercise 9

1. Distinguish between the following entities:
   (a) The CNS and the PNS
   (b) The efferent nervous system and the afferent nervous system
   (c) The autonomic NS and the somatic NS
   (d) The sympathetic NS and the parasympathetic NS
2. Describe the role of the sympathetic division of the ANS and its effects on the body.
3. Draw a simple diagram of CNS meninges consisting of three concentric circles. Label the three meninges, the epidural space, the sub-arachnoid space and the CSF. Further draw two needles: one delivering an epidural anaesthetic and the other in place for a lumbar puncture to draw out CSF.
4. Distinguish between a unipolar and a multipolar neuron.
5. (a) What is a "graded potential"?
   (b) What is an "action potential"?
6. Name four neurotransmitters. What is their function?
7. (a) What are the five main structures of the CNS?

(b) In which of the five are the following structures/areas located: conus medullaris, respiratory centres, motor areas, cauda equina, autonomic control centre, visual association areas, area for the control of body temperature, cervical enlargement, arbor vitae, basal nuclei, substantia nigra and posterior median sulcus?

8. What constitutes a reflex arc? Describe the stretch reflex.
9. What type of chemical is stored in the synaptic vesicles of a neuron?
10. Draw diagrams that distinguish between a unipolar, a bipolar and a multipolar neuron.
11. Draw a cross-sectional diagram of the spinal cord and label it with the structures: spinal meninges, grey matter, white matter, dura mater, arachnoid, pia mater, dorsal (posterior) grey horns, ventral (anterior) grey horns, sensory fibres, dorsal (posterior) root ganglion, spinal nerves, grey commissure, lateral grey horns, central canal, anterior median fissure and posterior median sulcus.
12. What is the cauda equina?
13. What is the difference between the spinal cord at L1 and L3?
14. Which two meninges enclose the CSF?
15. What is the name of the "space" between meninges that contains CSF?
16. If a patient has raised intracranial pressure, would CSF pressure in the lumbar region (measured by spinal tap) be higher or lower than normal? (Think of Pascal's principle.)
17. What structures separate the epidural space from the sub-arachnoid space?
18. What major nerves pass through the epidural space?
19. Where are the cell bodies of peripheral sensory neurons located?
20. Where are the cell bodies of peripheral motor neurons located?

Lecture 17: Special Senses

(Caon & Hickman 3rd ed, Ch11, pp. 297–300, Ch 12, pp 314–320)

1 General Senses

Sensory receptors are located throughout the body. They may be free nerve endings or nerve endings surrounded by a capsule.

Free nerve endings (including the Merkel disc in the skin and hair follicles) respond to cold, heat, vibration, pressure, touch, itch, pain and stretch.

Encapsulated nerve endings include Meissner's corpuscles (in hairless skin), Pacinian corpuscles (in skin, around the bones and joints), Ruffini's corpuscles (dermis, hypodermis, joints), muscle spindles (skeletal muscle) and Golgi tendon organs (tendons).

2 The Eye

Our eye is the structure that detects visible em radiation (light). A benefit of having two eyes is stereoscopic vision (we perceive three dimensions), so we can estimate distances and the position of objects in space.

- Fibrous tunic (sclera, cornea), aqueous humour (in the anterior cavity, between the lens and cornea)
- Vascular tunic (choroid, ciliary body and suspensory ligament, pupil, iris, lens), vitreous humour (posterior cavity, post. to lens)
- Sensory tunic (retina = pigmented layer, rods and cones, bipolar cells, ganglion cells), macula lutea, fovea centralis (cones), except at blind spot (optic disc)

(Light photons traverse the cornea, aqueous humour, pupil, lens, vitreous humour, axons, ganglion cells, amacrine cells, bipolar cells and horizontal cells to strike the rods and cones.)

M. Caon, *Lecture Notes, Worksheets, and Exercises for Basic Anatomy and Physiology*, https://doi.org/10.1007/978-3-031-56296-9_10

Light enters the posterior chamber of the eye through the black **pupil**, then passes through the lens and then traverses the vitreous. The diameter of the pupil may be varied by the colourful **iris** (a sphincter muscle under autonomic nervous system (ANS) control), from about 3 mm in diameter in bright light to about 8 mm in the dark. This change will vary the amount of light entering the eye by a factor of 7; the retina is able to adapt to different lighting conditions also (the anaxonic horizontal cells and amacrine cells of the retina also have a role in the eye's adjustment to dim or bright conditions).

The **retina** is the light-sensitive part of the eye. It converts light energy into electrical nerve impulses, which are sent to the brain. The retina covers the back half of the eyeball, enabling wide-angle vision.

Most vision is restricted to a small area called the **macula lutea** (yellow spot) and all detailed vision to a very small area called the **fovea centralis** (0.4 mm in diameter). At the fovea, the several layers of nerve tissue that overlie the retina are pushed aside so light strikes the cones directly (without having to traverse axons, ganglion cells and bipolar cells).

Rods and **cones** are the two types of photoreceptor cells in the retina.

The cones are used mainly for daylight (photopic) vision and colour vision and are primarily found in the fovea (but occur throughout the retina); each cone <u>in the fovea</u> has its own nerve connection to the brain (hence, vision is acute here), while in the rest of the retina, several cones share one nerve fibre.

The rods are used for low light (scotopic) vision (e.g. at night) and peripheral vision. They cover most of the retina and are much more numerous than cones. Hundreds of rods share the same nerve fibre, so rods have a poor ability to resolve close sources of light.

The **blind spot** on the retina is without rods or cones. Here, the blood vessels and axons of ganglion cells come together as the **optic nerve** and leave the eye at the "optic disc".

Visual Pathway

Axons from retinal ganglion cells, issue from each eye as **optic nerves (II)**. At the **optic chiasma**, fibres in the left (L) and right (R) optic nerves from the <u>medial aspect</u> of each eye cross over and continue (along with those from the lateral aspect of the contralateral eye) as L and R optic tracts and synapse with neurons in the **lateral geniculate body** (of the thalamus).

(Superior colliculi are involved in pupillary and eye movement reflexes via collateral connection.)

Many axons then continue as **optic radiation** of fibres into the **primary visual cortex** (in the occipital lobes).

(Note: Images of objects to our right fall onto the left half of the retina, and nerves from the left side of *each* eyeball go to the left side of the brain.)

Lenses

A **lens** is any transparent object with curved faces. If the curve is "outwards", making the middle of the lens fat and the edges thin, the lens is **convex**; the opposite curve is **concave**.

The focal length of a lens is the closest distance to the lens that an image can be formed (~2 cm for our eye).

Our eye can be thought of as a two-lens system. Both the cornea and the eye lens have curved faces, so both refract (and aid in focussing) light that enters our eye.

Accommodation

This refers to our ability to alter our eye's lens so that it can focus on objects at any distance away.

A thick, highly curved lens has a short focal length (for focussing light diverging from close objects); a thin, slightly curved lens has a long focal length (for focussing parallel light rays from distant objects).

Our lens is a "fat" biconvex shape, flexible and elastic, so it can be stretched thinly (by relaxing, the ciliary muscle retreats from the lens into a large circle, which pulls on the ciliary fibres, causing them to stretch the lens). This brings distant objects into focus.

Our lens can elastically recoil into a rounder, thick lens (by contracting ciliary muscles into a small circle, which releases the tension on ciliary fibres, allowing the lens to "ooze" back to shape). This brings close objects into focus.

Defects of Vision

Myopia – nearsightedness (too much refraction); it can be corrected with a concave lens.

Hyperopia – long-sightedness (too little refraction); it can be corrected with a convex lens.

Astigmatism – the cornea is unevenly curved, i.e. non-spherical.

Presbyopia (old-age vision) is caused by the lens losing its elasticity, so it cannot ooze into the short focal lengths of a youthful lens.

Corneal transplant: there is no rejection problem as there are no blood vessels to carry white blood cells (WBC).

Age-related macular degeneration: this refers to a blurred area near the centre of vision (the macula lutea).

"**Colour blindness**": this happens when three pigments in the cones are sensitive to red, green or blue light and also have some sensitivity beyond their major colour.

If the red-sensitive cones are missing, then red is perceived as green, and the condition is called **protanopia**.

In the reverse situation, green cones are missing, so green light stimulates red cones only, a condition called **deuteranopia.**

Anomalous trichromats: in people with this condition, all of their three cone types are used to perceive light colours, but one type of cone perceives light slightly differently.

Cataracts: there is opacity in the lens. The lens is removed (by phacoemulsification) and replaced with a soft plastic trifocal intraocular lens (IOL).

Other eye problems include glaucoma, a retinal bleed, retinal detachment

3 Ear Structure

Outer ear: auricle, external auditory meatus, tympanic membrane
Middle ear (filled with air): eardrum, tensor tympani, malleus, incus, stapes, stapedius muscle, oval window (OW), Eustachian tube
Inner ear: vestibule, cochlea, semi-circular canals, endolymph

The Transmission of Sound Waves to the Auditory Nerve
The ear is an "impedance matching" device, thereby directing sound propagating through air, via the middle ear, into the liquid endolymph of the cochlea.

(Resonance in the ear canal, leverage by ossicles, area ratio of eardrum and OW)

The **tympanic membrane** (TM, eardrum) separates air of the external ear canal from air in the middle ear (air in the middle ear is vented to the atmosphere through the **Eustachian tube**). TM vibrates due to pressure variations associated with the compressions and rarefactions of sound waves. TM moves the **malleus**, which moves the **incus**, which moves the **stapes** (the three ossicles), which presses on the **oval window** (of the cochlea). Hence, vibrations of air pass through OW into the fluid of the **cochlea** (the snail-like inner ear). Vibrations in the cochlear fluid cause movement in a certain part of the **basilar membrane** (on which sits the **organ of Corti**). The o. of C. is covered by the **tectorial membrane** so that its hair cells rub against the TM, which causes nerve impulses to be generated in the o. of C., which are transmitted to the brain via the cochlear nerve (part of Chapter VIII).

Sensitivity of Human Hearing

Sound intensity (in watts per square metre, W/m^2) is the objectively measurable amount of *sound energy* carried by a sound.

10^{-12} W/m^2 = silence (threshold of hearing).
1 W/m^2 = pain.

Sound level (in decibels (dB)) is a subjective measure of the loudness of a sound.

0 dB = silence.
0 dB does not mean that the amount of sound *energy* is zero. Just that we cannot hear it!

Two sounds that differ by 0.1 dB can barely be perceived as different in loudness; a sound that is 3 dB louder than another sounds twice as loud. Continuous noise at 90 dB will produce hearing damage without causing pain (→ encourage the wearing of ear-muffs – an occupational health requirement!)

The human ear is not equally sensitive to all audible frequencies.

Perceived loudness (in phons) – e.g. 60 dB "phon" (or equal loudness curve) is a graph of the sound level (in dB and over the range of audible frequencies) that we perceive to be the same loudness as a sound of 1000 Hz at 60 dB.

Audible range is ~20 Hz to ~20,000 Hz; infrasonic: <20 Hz, ultrasonic: >20 kHz.

The sounds of different frequencies stimulate our hearing in different amounts. The ear is most "sensitive" from 500 to 6000 Hz. It is not very sensitive below 200 Hz or above 12,000 Hz.

As we age, we progressively lose the ability to hear frequencies >~12,000 Hz (presbycusis).

For audible range test (Google "human audible range"). Try:

http://www.youtube.com/watch?v=qNf9nzvnd1k (continuous range)
http://www.youtube.com/watch?v=2G9Q-r2leyw (step changes)
http://www.youtube.com/watch?v=VxcbppCX6Rk (how old are you hearing test – wear headphones)

Other hearing problems include noise-induced hearing loss, Conductive deafness, Sensorineural loss.

4 The Sense of Equilibrium

(Involves input from the eyes, stretch receptors in muscles/tendons, and the vestibular apparatus)

The **vestibule** contains two membranous sacs (linked by a duct) called **saccule** and **utricle**. Each contains a sensory receptor called a **macula**, which is sensitive to *linear* acceleration forces.

The dense otoliths (embedded in the jelly-like otolithic membrane) have a relatively large inertia and, hence, respond more slowly than the body to changes in motion.

"Hairs cells" in the macula are bent (and stimulated) when the overlying **otolithic membrane**, in which they are embedded, moves.

The membrane in the saccule is vertical (responds to vertical acceleration). The membrane in the utricle is horizontal (responds to horizontal acceleration).

The three semi-circular canals each have an enlargement called an **ampulla**. The ampulla houses an equilibrium receptor called a **crista ampullaris**, with "hair cells" that respond to the rotation of the head. The endolymph in the ducts, because of inertia, remains stationary when the head (i.e. the "hair cells") begins to move (or the endolymph continues to move when the hair cells suddenly stop).

When these hair cells are moved through the stationary endolymph (or the moving endolymph moves over the stationary hair cells), they are stimulated.

The maculae and cristae ampullaris send impulses via the vestibular nerve (part of Chapter VIII).

Motion sickness is an equilibrium problem

5 Taste and Smell

Five basic tastes:

1. Sweet (sugars, alcohols, saccharins)
2. Sour (acids, lemon juice)
3. Salty (metal ions, esp. Na^+)
4. Bitter (quinine, caffeine, nicotine)
5. Umami (glutamate – beef taste, aged cheese, monosodium glutamate (MSG))

Also involved in "taste" are temperature, texture and pain, e.g. chilli. And "taste" is 80% smell!

Smell relies on volatile chemicals – molecules must enter the nose to be smelled. Olfactory receptors are neurons (replaced every ~60 days). Classifying "primary odours" is problematic.

6 Special Senses Revision: Homework Exercise 10

1. Describe the function of the (a) cornea, (b) lens, (c) retina, (d) canal of Schlemm, (e) choroid and (f) pupil.
2. Describe the role of the ciliary body and suspensory ligaments in the process known as accommodation.
3. Why does the retina have a blind spot?
4. What are the names commonly given to the two types of lenses?
5. Describe the effect on the light rays of each type of lens (diverge/converge).
6. Which one produces a magnified image when placed over some writing on a page?
7. Which one has the same curvature as your eye's lens?
8. Define what is meant by hyperopia.
9. Describe the function of the (a) pinna, (b) ossicles, (c) Eustachian tube and (d) cochlea.
10. What are the average sound frequency endpoints of the audible range for humans?

Lecture 18: Blood

Blood is the complex liquid component of the cardiovascular system (CVS). It is a connective tissue.

1 Functions

(a) Transport: oxygen (**O_2**) from lungs to cells, carbon dioxide (**CO_2**) from cells to lungs, **nitrogenous waste** from cells to kidneys and sweat glands, **nutrients** from digestive organs to the liver and cells, and **hormones** from the endocrine system to target cells and enzymes
    (Plasma proteins transport lipids, drugs etc.)
(b) Regulation: temperature (transfers heat around the body), pH (through the use of buffers), fluid volume and electrolyte balance
(c) Protection: coagulation; antibodies, white blood cells (WBC) and complement proteins defend against toxins and foreign microbes
    (Hydrostatic function in reproduction!)

Blood volume is ~8% of body mass. Hence, an average male has 5-6 litres and a female 4-5 litres. It is 45% **formed elements** (cells and cell fragments such as platelets) and 55% straw-coloured **plasma**.

Serum = plasma minus fibrinogen, Ca^{++} (i.e. the clotting factors).

Plasma

Its composition is 90% water and 8% protein (albumins 58%, globulins 37% (α, β, γ), fibrinogen 4%, peptide hormones, prothrombin, transferrin, plasminogen, clotting factors). The protein component helps maintain blood volume due to **colloid osmotic pressure** (and they act in clotting and transport lipids and iron).

Plasma also contains dissolved ions, wastes, dissolved gas, glucose & cholesterol.

M. Caon, *Lecture Notes, Worksheets, and Exercises for Basic Anatomy and Physiology*, https://doi.org/10.1007/978-3-031-56296-9_11

Formed Elements

– Made in red bone marrow

Haemopoiesis: haemocytoblasts →

→ Proerythroblast for red blood cells (RBC)
→ Lymphoid stem cell for lymphocytes
→ Myeloblast (and monoblast) for "….ophils" (and monocytes)
→ Megakaryoblast for platelets

RBC (erythrocytes) are 7–8 μm in diameter, have no nucleus (or mitochondria, endoplasmic reticulum) and contain haemoglobin (1/3 of the weight of the cell) to carry O_2. Of cells, 99.9% are RBC ($5 \times 10^6/\mu L$).

Their bi-concave shape allows distortion (folding) at constant volume without tearing.

WBC (leucocytes) have a nucleus and do not contain haemoglobin. They function as a body defence mechanism: "specific" (NK, T and B lymphocytes) and "non-specific" (macrophages, microphages).

* Leucocytes can migrate out of the blood by squeezing between endothelial cells (diapedesis). They can also squish up to pass through capillaries.
* They are all capable of amoeboid movement.
* They are attracted to specific chemical stimuli.

Five Types

- 50–70% are neutrophils (microphages – phagocytes of bacteria).
- 25% are *lymphocytes* (they recognise specific "non-self" antigens ➔ T cells migrate to targets; B cells become plasma cells, which produce antibodies; and NK cells destroy abnormal cells – e.g. cancerous ones).
- <8% are monocytes (➔ macrophages and osteoclasts). They attract fibroblasts to form scars.
- <4% are eosinophils (microphages – attack parasitic worms by exocytosis).
- <1% consists of basophils (they release histamine to promote vasodilation and inflammation, as well as heparin and chemicals to attract other WBC).

See http://www.biochemweb.org/neutrophil.shtml for neutrophil "chasing" a bacterium.

Neutropenia =
Granulocytopenia =
Leukopenia =
Lymphocytosis =

Platelets (Thrombocytes) (Blood Clotting)

– Size is 1 × 4 μm
– Circulate for 9–12 days before removal by spleen phagocytes
– 150,000–400,000 per μL of blood
– Thrombocytopenia < 80 000/μL
– Thrombocytosis = excessive # platelets

Haemocytoblasts undergo differentiation into the fourth immature cell line, the megakaryoblast, and transform into a large cell, which sheds fragments of cytoplasm surrounded by a membrane.

These non-nucleated cell fragments are **thrombocytes** or **platelets**, involved in haemostasis (the prevention of bleeding).

Platelets are non-sticky to normal endothelium.

Platelets adhere to the collagen of damaged endothelial cells.

2 Haemostasis

Haemostasis = prevention of blood loss.

Holes in damaged vessels must be plugged very rapidly.

Blood Clotting

1. A cascade of reactions resulting in the activation of the enzyme (thrombin) that forms the clot.
2. A mechanism of clot formation by that enzyme.

All necessary inactive precursor proteins are present in the blood.

Thrombin is the enzyme that converts fibrinogen to an insoluble fibrin clot (thrombus formation).

(a) THE VASCULAR PHASE: almost immediately (2 s) after a blood vessel is cut, the vessel walls **contract in a spasm** to slow the flow of blood (the vessel diameter decreases). Endothelial cells release chemical factors and local hormones. Endothelial cells become sticky.

(b) THE PLATELET PHASE: platelets *adhere* to exposed collagen fibres from the damaged vessel wall, to endothelial cells and to the basal lamina and release substances that cause other platelets to *aggregate* (+ve feedback), forming a platelet plug (within 15 s). Platelets release various compounds. The platelet plug is limited to the local area.

(Aspirin inhibits the production of factors causing platelet aggregation and is used in low doses in people susceptible to inappropriate clotting, which can lead to strokes and heart attacks.)

(c) **The Coagulation Phase**

Over 50 substances in the blood affect coagulation!

Clotting begins <2 mins from vessel damage.

Clot fills vessel opening in 3–6 mins.

Clot retracts after 20 mins (torn edges of vessel pulled together, reducing bleeding and the area of damage).

Clot changes to connective tissue over 1–2 weeks.

Three-Step Process of Blood Clotting (Coagulation)

1. **Formation of "Prothrombinase" (= *Pro*thrombin Activator) from Factor X**

This "first step" is actually the end result of a *series* of reactions from *two* pathways: the extrinsic pathway and the intrinsic pathway.

(In the **extrinsic** pathway, factor III combines with factor VII to form an "enzyme complex" that activates factor X.

In the **intrinsic** pathway, factor VIII combines with factor IX to form an "enzyme complex" that activates factor X.)

Extrinsic Pathway

It is called "extrinsic" because the substance (called tissue factor III) that initiates the formation of prothrombinase is released from damaged vessels (i.e. external to the blood).

It occurs rapidly (within seconds).

(Then prothrombinase converts *pro*thrombin to thrombin.)

Intrinsic Pathway

It is more complex and occurs more slowly (several minutes).

Damaged tissues cause platelet damage, and the platelets (internal to the blood) release phospholipids, which combine with various "clotting factors" to produce prothrombinase.

(Then prothrombinase converts *pro*thrombin to thrombin.)

2. **Formation of Thrombin**

Prothrombinase converts prothrombin to thrombin.

Thrombin formation has positive feedback, causing platelet aggregation and increased phospholipid release, leading to the formation of more prothrombinase (and therefore more thrombin etc).

Vitamin K is essential for prothrombin synthesis by the liver. Ca^{++} is essential for prothrombin conversion.

3. **Formation of Fibrin**

Thrombin acts as an enzyme to convert plasma fibrin*ogen* to insoluble fibrin polymer.

Fibrin forms a network of fibres, covering the platelet plug, which traps blood cells and fragments to form a clot.

(Fibrin – a monomer – can polymerise to form a "soft clot"; then cross-linking between fibrins produces a stable, web-like "hard clot" that entangles with RBC.)

3 Lysis of Clots (Fibrinolysis)

As repair proceeds, the clot gradually dissolves.

Plasmin*ogen* is converted to plasmin (by a tissue plasminogen activator (t-PA)). Plasmin destroys fibrin.

Summary of Coagulation

1. Factor XII (and platelet factor 3 from platelets) begins a cascade of reactions to form an enzyme complex (intrinsic p).
2. Tissue factor III (from damaged endothelial cells) begins a cascade of reactions to form an enzyme complex (extrinsic p).
3. Inactive factor X (is converted to -->) active factor X (by an enzyme complex).
4. Activated factor X binds to activated factor V to form the "prothrombinase complex"
5. Prothrombin (is converted to -->) thrombin (by prothrombinase).
6. Fibrinogen (is converted to -->) fibrin (by thrombin).
7. Plasminogen (is converted to -->) plasmin (by a tissue plasminogen activator).
8. Plasmin digests fibrin.

Ca^{++} and vitamin K are required for nearly every step in the clotting cascade (we always have enough Ca^{++}).

About half of our vitamin K is produced by intestinal bacteria and half from diet. Vitamin K is a fat-soluble vitamin essential to prothrombin synthesis in the liver (and for three other factors).

People suffering from liver disease (cholecystitis, hepatitis, cirrhosis) may experience uncontrolled bleeding.

Disorders that prevent fat absorption may result in vitamin K deficiency.

Injected vitamin K should be given 4–8 h pre-operation.

Haemophilia

Thrombocytopenia is a condition where the amount of platelets are less than the healthy range

4 Anticoagulants

Two anticoagulants used extensively in the clinical setting are heparin and warfarin.

Heparin

- Given intravenously
- Produced by basophils and mast cells (extracted from the tissue of cattle and pigs)
- Prevents the conversion of prothrombin to thrombin
- Involves rapid action and so is used extensively post-op (open heart surgery, bypass, valve replacement) and to prevent further thrombus formation in stroke victims
- Antidote = protamine sulphate (IV)

Warfarin

- Coumarins, rat poison
- Given orally
- Antagonist to vitamin K
- Involves slower acting, often taking days to have a therapeutic effect
- Antidote = vitamin K administration (IV)

Heparin must be administered intravenously, while warfarin can be taken orally. A standard anticoagulation regime would involve immediate IV heparin administration continued while warfarin builds up to therapeutic levels, allowing the patient to eventually go home on oral warfarin alone.

Purpura – purple spot areas of spontaneous bleeding into the skin due to platelet and vascular defects

Petechiae – small, punctuate skin haemorrhages; pinpoint, purplish red spots

Ecchymoses – bruises

Idiopathic thrombocytopenic purpura – an autoimmune disorder where antibodies to platelets form ➔ excess destruction of platelets

Splenomegaly occurs in cirrhosis with portal hypertension, as well as lymphomas ➔ an enlarged spleen captures and sequesters 80% of platelets.

Phlebitis – inflammation of a vein, often in the leg

Chronic venous hypertension (high blood pressure in the leg veins) may cause leg swelling, pain, varicose veins, ulcers.

5 Blood Groups

The surfaces of erythrocytes contain genetically determined glycoproteins called **antigens** (or **agglutinogens**).

1. **ABO System**

The ABO group is based on the presence or absence of two antigens on RBC: symbolised type A and type B. Your ABO blood group will be one of A, B, AB or O.

Circulating in the blood are pre-formed **antibodies** (= **agglutinins** = **immunoglobulin**) that recognise ABO antigens NOT present on your <u>own</u> RBC and attack introduced RBC, causing agglutination (clumping).

(A newborn lacks these antibodies, but these appear within a few months of birth,)

A person of blood group A has the A antigen on their RBC (consequently, their blood contains agglutinins against B).

A person of group AB has both A and B antigens on their RBC (consequently, their blood does NOT contain agglutinins against A or B).

A person of blood group O has neither the A nor B antigen on their RBC (consequently, their blood contains both agglutinins against A and B).

A person given incompatible blood will suffer a transfusion reaction.

The antibodies in the recipient's blood attack the foreign RBC, causing donated RBC to agglutinate.

Such intravascular clotting can lead to severe kidney damage.

2. **Rh System**

(It is so named because antigen D was first found in the blood of Rhesus monkeys.)

It has more than eight antigens on the RBC, the most important of which is antigen D (C and E are also common).

People with antigen D on their erythrocytes are called "Rh+" (81% of Australians) and those without "Rh-".

An Rh+ person (of course) does not contain antibodies against D.

An Rh- person also does not usually contain antibodies against D!

When an Rh- person receives Rh+ blood, their body will begin to produce agglutinins against the foreign Rh antigen D.

These agglutinins will remain in the blood.

If a second transfusion of Rh+ blood is given later, the agglutinins against D will now attack the donor erythrocytes, and a severe reaction may occur.

One of the most common problems with Rh incompatibility occurs during pregnancy if *the mother* is Rh- and the first baby is Rh+. During delivery, foetal blood may enter the maternal system from the placenta. The mother will then produce agglutinins against D. These can cross the placenta in a subsequent pregnancy, and if the second baby is Rh+, they will attack the foetal erythrocytes, causing **newborn haemolytic disease**.

Affected babies can have all their blood supply removed and replaced with Rh- blood. If identified, the disorder can be easily prevented by the administration of anti-D antibodies to the mother *after the first delivery* (17% of Australian mothers require these). These tie up the Rh+ erythrocytes from the foetus before the mother can start manufacturing her own agglutinins.

Blood/Blood Product Transfusions

Advances in transfusion technology allow for the separation of various blood components, and these can then be administered as needed.

Whole blood – used to increase blood volume, for example, after a haemorrhage

Packed cells – boost haemoglobin to restore O_2 carrying capacity without placing hypervolaemic stress on the system

Plasmapheresis: blood is removed, but RBC, WBC and platelets are returned to the donor. Plasma is used to make 17 products for various trauma, burns, cancer and blood disease treatments (see Red Cross at http://www.donateblood.com.au/all-about-blood/different-donation-types/plasma-products).

Blood group	Incidence in the USA West. Euro Africa		Antigen (on RBC)	Compatible recipient of	Agglutinins (in plasma)	Compatible donor to
A	40%	29%	A	A & O	Against-B	A & AB
B	11%	17%	B	B & O	Against-A	B & AB
AB	4%	4%	A & B	AB, A, B, O	neither	AB
O	45%	50%	neither	O	Against-A & Against-B	A, B, AB, O

Persons with blood group AB+ are "universal recipients" – will not agglutinate any ABO blood.

Persons with blood group O- are “universal donors” – have no antigens and so will provoke no agglutination.

There are eight different blood types. The graph shows the percentage of Australians who have a particular blood type.

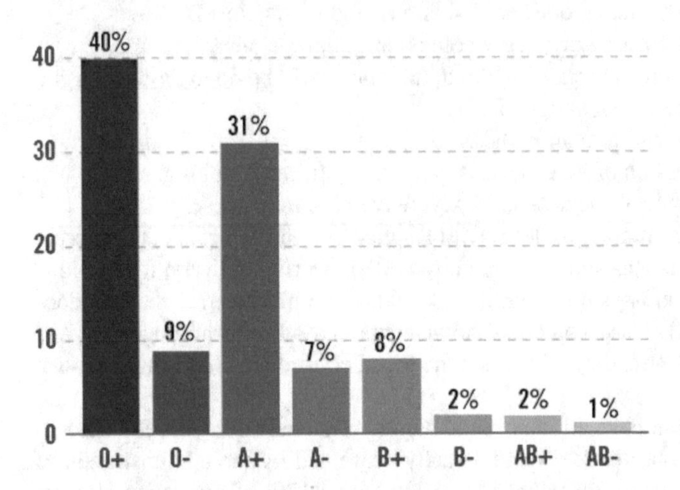

It is preferable for patients to receive blood transfusions of the same ABO and Rh(D) groups. However, in an emergency, if the required blood group is not available, a patient may be given another group as below.

Patient's Blood Type \ Donor's Blood Type	O-	O+	B-	B+	A-	A+	AB-	AB+
AB+	✓	✓	✓	✓	✓	✓	✓	✓
AB-	✓		✓		✓		✓	
A+	✓	✓			✓	✓		
A-	✓				✓			
B+	✓	✓	✓	✓				
B-	✓		✓					
O+	✓	✓						
O-	✓							

Spleen

- In the L hypochondriac region; the largest single mass of lymphoid tissue.
- Adjacent and inferior to the diaphragm, dorsal to the stomach and lateral to the L kidney.

- A soft textured, red organ about 12 cm long
- Contains “red pulp” and “white pulp”.
- Arterioles deliver blood to the red pulp, which is drained by sinusoids to venules.

Functions:

1. Phagocytosis of abnormal blood cells
2. Storing iron (Fe) recycled from RBC
3. Storing up to 1/3 of the body’s platelets
4. Initiating immune response by T and B cells against antigens in the blood (in the white pulp)
5. A major storage of monocytes; released into the blood after myocardial infarction (MI) to participate in healing

A splenic vein empties into the hepatic portal vein.

6 Additional Information

(CO_2 combines with the N-terminal groups on the four globin chains (not to the O_2 binding site). However, because of allosteric effects on the haemoglobin molecule, the binding of CO_2 decreases the amount of oxygen that is bound for a given partial pressure of oxygen.

The Haldane effect: the decreased binding to carbon dioxide in the blood due to increased oxygen levels; it is important in the transport of carbon dioxide from the tissues to the lungs.

The Bohr effect: a rise in the partial pressure of CO_2 or a lower pH will cause the offloading of oxygen from haemoglobin.

An **immunoglobulin** (Ig), also known as an antibody (Ab), is a large Y-shaped protein produced by B cells that is used by the immune system to identify and neutralise foreign objects such as bacteria and viruses. The antibody recognises a unique part of the foreign target, called an antigen.

Antibodies are secreted by a type of white blood cell, called a plasma cell. Antibodies can occur in two physical forms: a soluble form that is secreted from the cell and a membrane-bound form that is attached to the surface of a B cell and is referred to as the B cell receptor (BCR).

Blood Products from Plasma

(a) **Intramuscular immunoglobulin** (normal immunoglobulins – extracts from plasma that carry antibodies against common infectious diseases such as measles, rubella and hepatitis A)
(b) **Hepatitis B immunoglobulin,** Zoster immunoglobulin, **cytomegalovirus (CMV) immunoglobulin**, **tetanus immunoglobulin** (hyper immunoglobulins – prepared from a pool of donations from donors who have strong antibodies against diseases such as tetanus, chicken pox, hepatitis B and cytomegalovirus)

(c) **Rh(D) immunoglobulin** (anti-D – prevents Rhesus disease in newborns; Rhesus disease is an incompatibility of the mother's and baby's blood, resulting in the mother developing antibodies against her baby's blood, which can lead to the baby's red cells being destroyed. This disease has almost been eradicated, thanks to the availability of anti-D produced from selected blood donors.)
(d) **Factor VIII and von Willebrand factor (vWF)** (biostate (factor VIII concentrate) – a blood clotting factor used in the treatment of people with bleeding disorders such as haemophilia A or von Willebrand disease.)
(e) **Prothrombin complex concentrate** (prothrombinex is rich in coagulation factors II, IX and X and is used in the management of bleeding due to a deficiency of these proteins.)
(f) **Intravenous immunoglobulin IVIg** (Intragam – used to boost the immune system and in the treatment of some muscle and nerve conditions)
(g) **Albumin** (Albumex 20 is a concentrated solution of the main blood protein, albumin, present in human plasma. Albumex 20 is used in the correction of protein deficiency sometimes associated with kidney and liver diseases. Albumex 4 is a more dilute solution of albumin. It is used in the treatment of shock due to blood loss. It is also used in the treatment of shock after a person suffers severe burns.)

7 Blood Lecture Revision: Homework Exercise 11

1. Outline the functions of the blood and state which components (plasma, erythrocytes, leucocytes and platelets) are responsible for which properties of blood.
2. What is the structure, composition and function of RBC? And outline briefly the functions of white cells.
3. State briefly the effects of iron deficiency, amino acid deficiency and vitamin B12 deficiency on the blood. (Some research may be necessary!)
4. Outline the role of thrombocytes in haemostasis and describe the three main processes in coagulation, including the differences between extrinsic and intrinsic formation of prothrombin activators (exclude clotting factor names).
5. State the role of vitamin K in clotting and explain how uptake from the gut can be affected by disorders in fat absorption (fat-soluble enzyme).
6. Explain the clinical uses of heparin and warfarin in relation to their speed of effectiveness as anticoagulants.
7. Explain transfusion reactions by your knowledge of the ABO blood group and relate newborn haemolytic disease to the Rh blood group.
8. What are whole blood, packed cells and cryoprecipitate (clotting factors) used for in blood transfusions?
9. One microlitre (1 μL = 10-3 mm3) of blood contains about 5 million (5 × 106) RBC, and each RBC contains about 280 million (280 × 106) Hb molecules. Estimate how many Hb molecules there are in an adult human.
10. Name the three types of white blood cells depicted below.

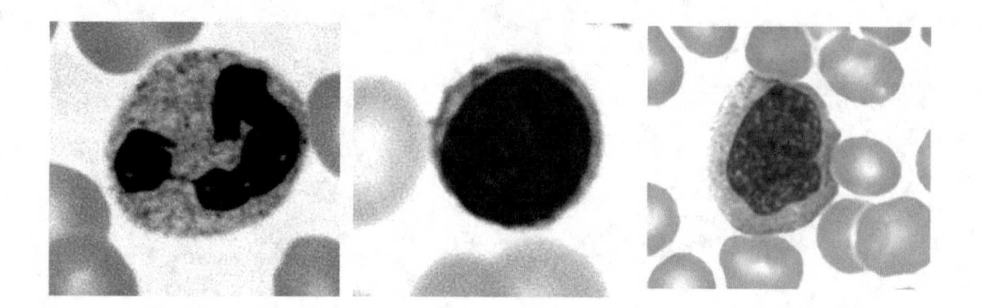

Lecture 19: Cardiovascular System: Anatomy of the Heart

The cardiovascular system consists of the heart (the muscular pump) and a closed system of arteries, capillaries and veins for the circulation of the blood.

The cardiovascular system is functional long before any other major organ system in the human embryo, and the primitive heart is beating regularly early in the fourth week following fertilisation.

1 Position

The heart is a four-chambered muscular organ roughly the size of a closed fist. It lies between the lungs in the mediastinum from the second to sixth ribs. About 2/3 of its mass lies on the left of the body's midline.

The base (broad end) of the heart consists mostly of the atria and is directed superior-dorsally and to the right. The apex (pointy end) is formed by the tip of the left ventricle and rests on the diaphragm pointing inferio-ventrally and to the left.

2 Covering

The heart is enclosed in a loose-fitting, two-layered sac called the **pericardium**. The outer layer, or fibrous pericardium, consists of tough, fibrous connective tissue. It is continuous with the connective tissue surrounding the great vessels that enter the heart. The inner layer, or the serous pericardium, consists of two membranes: a **parietal** membrane (a layer of thin, flattened cells called the mesothelium), which is fused to the inside of the fibrous pericardium, and a **visceral** membrane (also called the **epicardium**) adhering to the outside of the myocardium.

M. Caon, *Lecture Notes, Worksheets, and Exercises for Basic Anatomy and Physiology*, https://doi.org/10.1007/978-3-031-56296-9_12

Between the parietal and visceral membranes of the serous pericardium is a potential space, which usually contains a thin film of serous fluid for lubrication. Inflammation of the pericardium, called **pericarditis**, may occur following infection or myocardial infarction as a result of excess pericardial fluid, fibrin or pus accumulation. (A feature of pain diagnosis after a myocardial infarction is to determine if the pain is worse on inspiration. If so, it is likely to be due to pericarditis and not associated with further infarction and can best be relieved with aspirin.)

Accumulation of too much pericardial fluid can exert a compression on the heart called **cardiac tamponade**, possibly necessitating the removal of fluid from the pericardium or sometimes stripping the whole fibrous pericardium and its serous lining.

3 Heart Wall

It has three layers:

1. **Epicardium** or the outer layer is the same as the visceral membrane of the serous pericardium.
2. **Myocardium**: the bulk of the heart is a thick, contractile middle layer of specially arranged muscle cells called the myocardium. Cardiac muscle fibres are striated, uni-nucleate, intricately branched and joined to each other by "intercalated discs" (to form a *functional* syncytium). Skeletal muscle fibres are arranged in parallel (and **are** a syncytium!). Skeletal muscle is also classed as voluntary, while cardiac and smooth muscles are involuntary.
3. The myocardium is richly supplied throughout with nerves, blood vessels and lymphatic vessels to enable it to carry out its heavy workload.
4. **Endocardium** is the thin layer of epithelium lining the inside of the heart and valves and continuous with the endothelium of the large blood vessels.

4 Cardiac Tissue

Cardiac muscle cells are typically uninucleate and are 10–20 μm in diameter and 50–100 μm in length.

Twenty-five per cent (25%) of the cell volume is taken up by mitochondria!

Cells connected by porous junctions = **intercalated discs** (cell membranes of adjacent muscle cells extensively intertwined). They permit sodium, potassium and calcium to easily diffuse from cell to cell. This makes it easier for depolarisation and repolarisation in the myocardium. Because of these junctions and bridges, the heart muscle is able to act as a single coordinated unit. Cardiac muscle is a "functional syncytium".

Cardiac cells are held together by connections called **desmosomes** and communicate via "gap junctions" (interlocked membrane proteins that form a channel to allow the passage of ions and small molecules). Cardiac cells are also held together by myofibrils anchored to intercalated discs.

5 Chambers of the Heart

The interior of the heart is divided into four chambers. The upper two chambers are called the **atria** and are separated by the **interatrial septum**. The right atrium receives blood from the body via the **superior vena cava** (blood from the upper body) and **inferior vena cava** (blood from the lower body). The coronary sinus also empties into the right atrium and drains blood from the vessels supplying the heart. The interior of the right atrium has a depression called the **fossa ovalis**, which is the remnant of the opening in the foetal heart, called the foramen ovale.

The left atrium (LA) receives blood from the four pulmonary veins coming from the lungs. Each atrium has a pouched appendage called an auricle, which serves to increase the volume of the atrium. The right ventricle receives blood from the right atrium and pumps it via the pulmonary artery to the lungs.

The left ventricle (LV) has the thickest muscle walls as it must pump blood out through the aorta to the rest of the body. The right and left ventricles are separated by the interventricular septum.

6 Valves of the Heart

The heart has four valves (structures that ensure that blood flows in one direction only). They prevent blood re-entering the chamber that it just left.

Two atrioventricular (AV) valves: the **tricuspid** separates the right atrium and the right ventricle. The bicuspid (or **mitral**) lies between the left atrium and the left ventricle. (Remember alphabetically, L and M for left and mitral and R and T for right and tricuspid).

They are constructed of flaps of the endocardium and are anchored to the **papillary** muscles by chord-like structures called **chordae tendinae** (the AV valves are sometimes called "parachute valves" because of their appearance). As the ventricles contract, the papillary muscle contracts to tighten the chordae tendinae, thus preventing the flaps from moving too far. Hence, blood flows from the atria to the ventricles but is prevented from backflow into the atria.

The other two valves are called **semilunar valves** and are located at the start of the pulmonary and aortic arteries. They prevent the backflow of blood into the ventricles from these vessels when the ventricles relax.

Tough elastic tissue encircles the heart valves and the base of the aorta and pulmonary trunk. This is the **fibrous skeleton** of the heart. It electrically isolates ventricular cells from atrial cells.

7 Blood Flow Through the Heart

(From the upper body via) the superior vena cava and from the lower body via the inferior vena cava into the right atrium, through the tricuspid valve into the right ventricle; from RV through the pulmonary valve into the pulmonary trunk and pulmonary circulation; from the lungs (via two left and two right) into pulmonary veins into LA, then through the mitral valve into LV; and from LV through the aortic valve into the aorta and systemic circulation.

8 Blood Supply to the Heart

The walls of the heart, like any other tissue, need oxygen, nutrients and various other substances in order to survive. These are supplied via the **coronary circulation**. The left and right coronary arteries arise from the ascending aorta, immediately above the aortic valve, and branch out to supply blood to the myocardium. Blood enters the coronary arteries during ventricular systole.

(Left coronary ➔ circumflex and anterior interventricular arteries. Right coronary ➔ marginal(s) and posterior interventricular arteries)

Small branches of the left and right coronary arteries eventually join (anastomose) so that there is an alternate path for blood to a portion of the myocardium otherwise isolated by a vessel blockage.

Coronary artery disease, due principally to underlying atherosclerosis (build-up of fatty deposits, especially cholesterol, in the arteries), is a condition in which the heart muscle receives an inadequate amount of blood (becomes **ischemic**) because of an interruption of its blood supply. This may lead to **angina pectoris** (chest pain) due to the lack of oxygen to the part of the heart affected. Total occlusion of an arterial branch leads to the death of the myocardial tissue that it supplies and constitutes a **myocardial infarction**.

9 Cardiac Cycle

Cycle – the time period between the start of one heartbeat and the beginning of the next.

Ventricular **systole** (contraction of ventricles): the myocardium contracts; when pressure is high enough, blood is pushed through semilunar valves into the aorta/pulmonary trunks.

Ventricular **diastole** (relaxation of ventricles): blood in the aorta and the pulmonary trunk force semilunar valves shut, blood flows into atria and ventricles (through the AV valves).

10 Heart Sounds

The noise made by blood squirting through AV valves as they close = "lub", while the closing of semilunar valves = "dup".

Heart murmur = some blood "regurgitated" through the mitral valve back into the left atrium (not uncommon).

11 Conduction System

The conduction system of the heart – sinoatrial (SA) node, atrioventricular (AV) node, AV bundle, L and R bundle branches and the Purkinje system – consists of muscle tissue in which the cells have lost their ability to contract and have become specialised for generating and transmitting impulses that stimulate the cardiac muscle tissue to contract (see the physiology of the cardiovascular system).

12 Nerve Supply

The contraction of the atria and ventricles is initiated by the internal conduction system of the heart. However, the rate and force of contraction are largely under the control of the autonomic nervous system. **Sympathetic** nerve fibres from the cardio-accelerator centre in the medulla oblongata increase the rate and force of contraction. **Parasympathetic** connections from the cardio- inhibitor centre via the vagal nerve (CN. X) decrease cardiac activity (see blood pressure control later).

The 6-second electrocardiogram (ECG) simulator: https://www.skillstat.com/six-second-ecg-resources/

13 Electrocardiogram

Resting cardiac muscle cells have a negative voltage on the inside of their sarcolemma when compared to the outside of the sarcolemma. In this state, they are said to be "polarised".

The ECG is a graph of voltage generated by the heart versus time – as the heart is stimulated to contract, measured at the body surface.

The significant features (movements away from the horizontal) of this graph are labelled with the letters P, Q, R, S and T.

The heart generates a measurable voltage (electrical potential) when the cells of the atria electrically depolarise, when the atria repolarise, when the cells of the ventricles electrically depolarise and when they repolarise.

Atrial depolarisation produces the “P wave”.
Ventricular depolarisation produces the “QRS wave”.
Ventricular repolarisation produces the “T wave”.

Soon after depolarisation, the atria and ventricles contract (ventricular systole).
Soon after repolarisation, the atria and ventricles relax (ventricular diastole).

14 Additional Information

The valve of the coronary sinus (**Thebesian** valve) is a semi-circular fold of the lining membrane of the right atrium, at the orifice of the coronary sinus. It is situated at the base of the inferior vena cava. The valve may vary in size or be completely absent.

The **Vieussens** valve of the coronary sinus is a valve and anatomic landmark between the coronary sinus and the great cardiac vein. It is often a flimsy valve composed of one to three leaflets found in 80–90% of people.

Other valves exist in the posterior veins of LV (e.g.) often at venous branches.

The anterior cardiac veins do not drain into the coronary sinus but drain directly into the right atrium. Some small veins, known as the smallest cardiac veins, drain directly into any of the four chambers of the heart(!)

A Normal ECG Trace

P Wave: Atrial Depolarisation
This represents The P wave is defined as the electrical activity associated with the original impulse from the S-A node and its subsequent spread through the atria – in other words, depolarisation of the atrial section of the myocardium. If P waves are present, it can be assumed that the stimulus began in the S-A node. Following depolarisation, the atria must repolarise, but this occurs during the QRS complex and cannot be identified while the ventricles are repolarising.

P-R Interval:
Passage of Impulse from Atria to Ventricles. The period from the beginning of the P wave to the beginning of the QRS complex is called the P-R interval. It normally takes no more than 0.2 s and represents the time taken for the electrical impulse to pass from the right atrium to the ventricles and to initiate ventricular contraction.

QRS Complex: Ventricular Depolarisation
This is the largest feature of the ECG and generally consists of an initial downward deflection (Q wave), a larger upward deflection (R wave) and a second downward deflection (S wave). This complex represents the activation of the ventricles and is normally less than 0.12 s in duration. It should be noted that the relative size of the deflections in the QRS complex varies according to the lead from which the recording has been taken.

ST Segment
The period between the completion of depolarisation and the beginning of repolarisation of the ventricular muscle is represented by the segment from the completion of the QRS complex to the beginning of the T wave. This segment may be elevated above the zero line or depressed if there is acute cardiac muscle injury, as in a myocardial infarction.

T Wave: Ventricular Repolarisation
This wave represents the recovery phase (repolarisation of the heart muscle) after ventricular depolarisation. It is usually of longer duration and somewhat larger than the P wave. The T wave is variable, being influenced by physiological as well as pathological conditions; for example, raised serum potassium levels cause tall, tented T waves, while low K+ levels cause depressed T waves. If the myocardium is damaged in some way, e.g. from tissue injury or ischaemia, the T wave may be flattened or inverted.

The passage of the P, QRS and T waves represents one cardiac cycle.

15 Heart Lecture Revision: Homework Exercise 12

1. Which heart valve is also known as the tricuspid? (Which other heart valves also have three flaps?)
2. Which of (ventricular) systole or diastole refers to the contraction of the heart muscle?
3. Apart from the heart, what other structures lie within the mediastinum?
4. Arrange (write down) the four chambers of the heart (LA, RA, LV, RV), the four valves (Ra-v, La-v, Pul v, Ao v), the blood vessels (aorta, pulmonary trunk, pulmonary veins, vena cavae) in the order in which blood flows through them.
5. Name the membrane that adheres to the outside of the myocardium.
6. Distinguish cardiac muscle from skeletal muscle.
7. Where are the coronary arteries, and what do they do?
8. What do the heart valves do as the ventricles progress through systole?
9. Why is the myocardium of the left ventricle so thick?
10. On the schematic (i.e. not anatomically faithful) diagram of the heart below, draw in and label the aortic and pulmonary valves and label all the structures listed in question 4. In addition, use arrows to indicate the flow of blood.

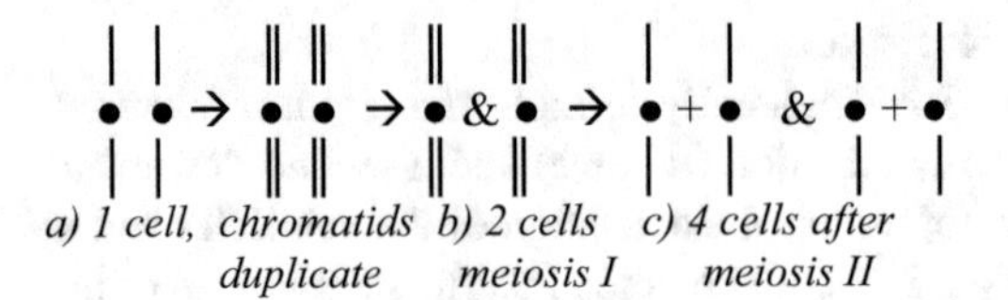

11. What is the general function of the heart valves, and what would happen to the circulation of blood through the heart if the action of the valves was impaired?
12. How does the structure of the left ventricle wall differ from that of the right?
13. How do these differences in ventricle walls relate to resistance to blood flow in pulmonary and systemic circulations?
14. Describe how and during which part of the heart cycle blood enters the coronary arteries.
15. What effect would a blockage of a coronary artery have on the myocardium (heart muscle)?
16. How many cusps does the RIGHT atrioventricular valve have? What is the other name for this valve?
17. How many cusps does the LEFT AV valve have? What is the other name for this valve?

Lecture 20: Anatomy, Structure and Function of Blood Vessels

1 Blood Vessels

There are three types of blood vessels: arteries, veins and capillaries.

Artery: a vessel that carries blood away from the heart. After birth, all arteries, except the pulmonary artery and branches, carry oxygenated blood. A small artery (diameter of 30 μm) is called an **arteriole**.

Vein: a vessel that carries blood towards the heart. All veins except the 4 pulmonary veins contain deoxygenated blood. A small vein is called a **venule**.

Capillaries: short (1 mm) microscopic vessels (diam 8 μm) that carry blood from the arterioles to the venules. The exchange through vessel walls occurs only in the capillaries. (William Harvey (c1630) first proposed that blood circulated from the heart through arteries and veins back to the heart. This hypothesis was rejected by many on the basis that there was no way for blood to get from the arteries to the veins. Only the advent of the microscope revealed the capillaries connecting arteries and veins.)

2 Arteries

They are composed of three coats or "**tunics**" and a central lumen:

1. Tunica intima – inner coat or wall composed of simple squamous epithelial tissue (endothelium)
2. Tunica media – the middle and usually thickest layer, composed of elastic fibres and smooth muscle; sympathetic NS innervation that causes vasoconstriction and vasodilation

M. Caon, *Lecture Notes, Worksheets, and Exercises for Basic Anatomy and Physiology*, https://doi.org/10.1007/978-3-031-56296-9_13

3. Tunica externa (adventitia) – an outer coat of white fibrous tissue (collagen and elastic fibres) that causes the artery to stand open instead of collapsing when cut

The largest arteries (up to diam 2.5 cm) are called **conducting arteries** (or elastic arteries) such as the aorta first brachiocephalic (→ R. common carotid, R. subclavian, R. vertebral), second L. common carotid and third L. subclavian (→ L. vertebral).

The descending aorta bifurcates into L & R common iliac arteries. Their tunica media contains more elastic than muscle tissue.

The extra pliability of the conducting arteries causes them to bulge when the heart contracts and forces blood into the aorta, thus accommodating the surge in blood and storing some systolic pressure as elastic potential energy. As the heart rests, the arteries recoil, moving the blood forward in a more continuous flow than could occur through rigid vessels. The alternate expansion and contraction of an artery constitute the arterial pulse.

Medium-sized arteries (diam 0.4 cm) are called **muscular arteries** (or distributing arteries) and include the axillary, brachial, radial, femoral, popliteal and tibial arteries. Their tunica media contains more smooth muscle than elastic tissue, enabling greater autonomic control of vasoconstriction and vasodilation.

About 15% of the blood volume is in arteries at any one moment.

3 Veins

Veins have the same three tunics as arteries, but since they do not have to cope with the same high pressures as arteries, the tunics are thinner and with less elastic tissue and smooth muscle.

Two common iliac veins join to form the inferior vena cava.

R internal jugular + external jugular + vertebral + subclavian veins join to form R brachiocephalic (same on the left). The two b–c join into the superior vena cava.

Factors Assisting Venous Return

Most larger veins, especially those in the limbs, have valves to prevent the backflow of blood under the influence of gravity. Venous return is assisted by skeletal muscle contraction. On contraction, muscles press against the blood in the veins, forcing the valves open and driving blood towards the heart (much as toothpaste is squashed out of the tube). (On long aeroplane flights, people often complain of feet swelling and being unable to put their shoes back on. This is a direct result of compromised venous return due to the skeletal muscles being immobilised.)

Breathing assists venous return since the downward movement of the diaphragm during inspiration causes an increase in abdominal pressure and a decrease in thoracic pressure. Thus, blood is forced up from the abdominal to the thoracic veins along the pressure gradient.

Suction effect of the heart: during early systole, the atrioventricular (AV) ring moves downwards, which expands the volume of the atria, creating a negative pressure.

Veins as Blood Reservoirs Assisting Blood Distribution
The veins also constitute a large blood reservoir (59% of the total blood volume), which can be manipulated via the vasomotor control mechanism to influence the distribution of blood within the system (see later section on blood pressure). (For example, if we suddenly require strenuous muscular activity, the vasomotor centre in the brain increases the sympathetic impulses to the veins in the intestine, liver, spleen and skin, causing vasoconstriction and diverting blood to the muscles and heart.)

4 Capillaries

The primary function of the capillaries is to permit the exchange of dissolved gases, nutrients, hormones and waste products between the interstitial fluid and the blood. Therefore, as one might expect, *the walls are extremely thin (0.5 μm) and composed of just a single layer of endothelial cells (**tunica intima**)* surrounded by a basement membrane.

Diffusion Across the Capillary Wall
Lipid-soluble substances (including oxygen (O_2) and carbon dioxide (CO_2)) diffuse through the plasma membrane.

Water-filled channels in the plasma membrane allow ions and polar molecules to diffuse through.

Intercellular clefts allow molecules to pass into and out of the capillary.

Endocytosis (via endosomes), followed by exocytosis, allows a small amount of protein to exit.

Capillaries are found in close proximity to most cells and are in the highest concentration in those areas where activity is greatest (muscles, liver, kidneys, lungs, nervous system).

Continuous capillaries (most common): the endothelium forms a complete lining. The cells are joined by "tight junctions" interrupted by "intercellular clefts" (except in the blood-brain barrier!). Allow the diffusion of water, small solutes and lipid-soluble materials, e.g. skin, muscles, lungs.

Fenestrated capillaries: the endothelium has pores (windows) to allow the rapid movement of water and solutes (incorporated peptides) between the plasma and interstitial fluid, e.g. villi of the small intestine, endocrine organs and glomerulus of the kidney.

Sinusoids appear as leaky "fenestrated capillaries" with gaps between adjacent endothelial cells (and little or no basement membrane). They allow the free exchange of water, plasma proteins, blood cells and phagocytic cells (in the liver, the bone marrow, the spleen and some endocrine organs).

Capillary beds (or plexus): dozens of capillaries may arise from one arteriole. Hence, blood flow is the slowest in the capillaries (about 1 mm/s), and this allows for the exchange of materials between the capillary blood and the interstitial fluid.

Entrance to each capillary is governed by a **pre-capillary sphincter**. Contraction causes a decrease in blood flow.

Most substances are exchanged by diffusion; hence, cells must be in close proximity to capillaries (within 0.125 mm).

Nutrients and oxygen move from the capillaries (where they are in high concentration) via the interstitial fluid to the tissues (lower concentration). Similarly, waste products, due to metabolic activity, move from the tissues (high concentration) into the interstitial fluid and then the blood (or lymph system; see later), to be transported for disposal via the lungs, kidneys or sweat glands.

Capillaries, Starling's Law and Microcirculation

Differences between the pressure in the capillaries and that in the interstitial fluid, and also between the *osmotic pressure* of the blood (colloid osmotic pressure due to blood proteins) and that of the interstitial fluid, result in a small net movement of fluid and solutes from the capillaries to the interstitial fluid at the arterial end of the capillary.

At the venous end of the capillary, this net movement is reversed. However, not all fluid filtered out at one end filters back in at the other. Some enter the **lymphatic system** before an eventual return to the cardiovascular system. The state of near equilibrium, in which fluid leaving the capillaries at the arterial end is balanced by that entering at the venous end, plus that returned by lymphatics, is known as Starling's law of the capillaries.

This movement of fluid between blood (via the capillaries), the interstitial fluid, cells and the lymph constitutes **microcirculation** and is vital for maintaining the internal environment of the human body.

5 Circulatory Routes

In **systemic circulation**, oxygenated blood (bright red) circulates from the left ventricle out through the aorta to the arteries, then to arterioles and, finally, to the capillaries, to supply the tissues with nutrients and oxygenated blood.

Deoxygenated blood (dark red) continues along the capillaries into the venules to the veins and via the superior and inferior vena cava into the right atrium.

Deoxygenated blood is pumped into the **pulmonary circulation** by the right ventricle (RV) via the pulmonary artery to the lungs, where gaseous exchange occurs and oxygenated blood returns via the pulmonary veins (the only veins carrying oxygenated blood), to the left atrium and then again into the left ventricle to complete the cycle. Blood vessels are shorter and more distensible than systemic circulation ∴ lower pressure

Coronary Circulation

The right and left coronary arteries branch from the ascending aorta and carry oxygenated blood to the arterial system of the myocardium. Deoxygenated blood returns to the right atrium via the coronary sinus (and anterior cardiac veins).

6 Disorders Affecting Blood Vessels

Arteriosclerosis: this happens when there is a build-up of fatty deposits (atheromas or plaque), especially cholesterol, in the walls of the arteries. It may result from damage to the artery lining, possibly as a result of high blood pressure, smoking, diabetes and high-cholesterol diets. Complications include angina pectoris, myocardial infarction, peripheral vascular disease and strokes (probably as a result of emboli detaching from the atheromatous plaques.

Aneurysm: this refers to the weakened section of the arterial wall that bulges out, forming a balloon-like sac, *most commonly in the abdominal aorta and in the brain.* The contributing factors are atherosclerosis, trauma and congenital vessel defects. A ruptured aneurysm commonly causes massive haemorrhage and death.

Varicose veins: weakened venous valves allow blood to flow back under gravity into the distal parts of the vein. Repeated overloading in this way causes the walls to become permanently dilated, losing their elasticity. Haemorrhoids are varicosities in the anal area. Contributing factors may be heredity, mechanical stress (prolonged standing, pregnancy) or age.

Deep vein thrombosis (DVT): a blood clot forms in a deep vein of the lower leg or thigh. Deep veins pass through the centre of your leg and are surrounded by a layer of muscle. A pulmonary embolism happens when a piece of the blood clot from a DVT breaks off and travels to the lungs, where it blocks one of the blood vessels. This is serious and can be fatal.

Arteriosclerosis – a disease of the arteries where the wall is thickened and loses elasticity due to the build-up of atheromas (plaque).

Atherosclerosis – fatty deposits (cholesterol and Ca^{++}) in the tunica media; it is the most common form of arteriosclerosis.

Arteriolosclerosis – arteriosclerosis of the arterioles.

7 HLTH1004 Blood Vessels Revision: Homework Exercise 13

1. What are the three tunics that comprise the walls of veins and arteries?
2. Name some differences between veins and arteries.
3. What is the systemic circulation?
4. What is the difference between muscular and elastic arteries?
5. What is the function of the smooth muscle in blood vessel walls?
6. Name the three types of capillaries. What is the difference between them?
7. What would result if "Starling's law of the capillaries" does not hold?
8. (a) What causes blood to flow in the arterial side of the systemic circulation?
   (b) What causes venous return (blood flow) in the venous side of the circuit?

Lecture 21: Pressure and Fluid Dynamics

(Caon & Hickman 3rd ed. pp 234–242, pp. 266–278 & Qu 33–40)

1 Definition and Units for Pressure

Pressure is the force (in N) being exerted, divided by the area (in m^2) on which the force acts. Note that pressure is NOT a force (nor is it a length, though one unit is "mm Hg").

$$P = F / A \quad \text{or} \quad \text{Pressure} = \text{Force} \div \text{Area}$$

The unit is N/m^2 = Pascal (Pa).

(When performing CPR, why use the heel of the hand rather than the whole palm to press down on the sternum?)

One pascal is a small unit, so pressure is usually expressed in **kilo**pascals (kPa).

Other Units

- Millimetres of mercury (mm Hg) – for BP!
- Centimetres of water (cm H_2O)
- Standard atmospheres (atm.) – for hyperbaric!
- Bars (hence *baroreceptors and barometers)*
- Pounds per square inch (psi) – for car tyres!

Conversion Factors

1 kPa = 1000 Pa = 7.5 mm Hg = 10.23 cm H_2O

1 mm Hg = 0.133 kPa = 1.36 cm H_2O

$$\left(120/80\,\text{mm}\,Hg \equiv 16/10.6\,kPa\right)$$

M. Caon, *Lecture Notes, Worksheets, and Exercises for Basic Anatomy and Physiology*, https://doi.org/10.1007/978-3-031-56296-9_14

2 Atmospheric Pressure and Pressure Shown on the Gauge

The air around us (79% N_2, 21% O_2, 0.5% H_2O, 0.04% CO_2) exerts a pressure due to the weight of the atmosphere above us, called atmospheric pressure.

It was first measured by Torricelli using a mercury **barometer** (atm. P. was measured by how high – in mm – a column of mercury could be supported by the pressure of the atmosphere).

The standard atmospheric pressure on 1 m^2 of area at sea level is due to the weight of air in the column rising to the top of the atmosphere. The column contains about 10,100 kg of air (which has a weight of about 101,000 N), but the amount varies from day to day.

1 atm = 101,000 N/m^2 = 1.01×10^5 Pa
101 kPa = 1010 hPa = 1010 millibars 760 mm Hg = 1030 cm H_2O = 14.7 psi

Atmospheric pressure decreases with altitude above sea level. At 3000 m, atm. P = 70 kPa. Our middle ear contains air that is separated from the atmosphere by our eardrum but may be vented by the *Eustachian* tube. The middle ear is sensitive to rapid changes in air pressure.

Atmospheric pressure is used as a *relative zero*. Pressures greater than atm. press. are "**positive**". Those less than atm. Press. are "**negative**" (suction).

The readings on pressure gauges show pressure values above or below atm. press. (and are called "gauge pressure").

That is, a systole pressure of 120 mm Hg means 120 *above* atmospheric and is actually 120 + 760 = 880 mm Hg. And car tyre pressures are 192 kPa (28 psi) *greater than atm. press.*

3 Suction

Suction involves creating a negative pressure so that atmospheric pressure will push the fluid in the direction of the negative pressure (e.g. drinking with a straw, filling a syringe, wound drainage using a *Bellovac* or *RediVac* or *PortaVac* bottle, taking blood using a *Vacuette*).

4 Boyle's Law

"The pressure in a fixed amount of gas will increase as its volume decreases (pressure is inversely proportional to volume)."

$$P \propto 1/V$$

Examples are expanding the lungs to breathe, a syringe and other "bicycle pump" style devices.

5 Hydrostatic Pressure

Pressure in liquids increases with depth because the weight of liquid above increases with depth (about 1 atm. per 10 m).

Pressure due to the weight of liquid or the "**head**" of liquid above a point is called **hydrostatic pressure** (HP), stated in units "cm H_2O", e.g. pressure at the cannula of an intravenous (IV) saline infusion with a head of liquid of 40 cm

$$\begin{aligned} P &= 9.8 \times \text{density} \times \text{head} \\ &= 9.8 \times 1000 \ \text{kg/m}^3 \times 0.4 \ \text{m} \\ &= 3920 \ \text{Pa} = 3.9 \ kPa \ (= 29 \ \text{mm} \ Hg) \end{aligned}$$

This pressure must be greater than blood pressure for IV liquid to flow in.

While standing, there is a pressure difference between feet and head due to gravity acting on the "head of liquid". Hence, when standing, blood pressure in the feet will be increased by the head of the blood contained in the vessels (blood pressures in the body are stated as if the body is supine).

Note that:

1. At the surface of the liquid (where depth is zero), pressure is atmospheric.
2. Pressure difference, due to the head of liquid, depends only on the depth in the liquid, not on the amount of the liquid nor on the shape of the vessel.
3. The pressure at any point in a liquid that is at rest acts equally in all directions (otherwise, the liquid would not be at rest).
4. **PASCAL'S PRINCIPLE**: Because liquids are incompressible, "pressure applied to an enclosed liquid at rest is transmitted undiminished to every portion of the liquid and to the walls of the containing vessel".

(For example, relief of pressure sores; Queckenstedt's test of cerebrospinal fluid (CSF) pressure; squeezing an IV bag that cannot be hung up; increased eyeball pressure in glaucoma; distribution of pressure in the knee joint, thanks to enclosed synovial fluid; foetus enclosed by amniotic fluid)

Soon after the heart's ventricles fill with blood, they contract, and pressure on the blood in the left ventricle (LV) rises rapidly from about 0 mm Hg to 80 mm Hg. The atrioventricular (AV) valves (mitral and tricuspid) close immediately, but the semilunar aortic (and pulmonary) valves do not open until LVP = 80 mm Hg (and RVP = 8 mm Hg). That is, tension builds up in the cardiac muscle until the pressure exerted by the (incompressible) blood builds up enough to push open the semilunar valves (which are, until then, held shut by the pressure of the blood on the aorta side of the valve).

Pressure considerations in *flowing* fluids are slightly different than in *static* fluids.

6 Gases Dissolved in Water

Soluble gases like oxygen (O_2) and carbon dioxide (CO_2) dissolve in body fluids if the gas is in contact with the fluid.

The concentration of dissolved gas depends on the solubility of the gas and on its **partial pressure.**

Henry's Law

$$\text{Conc}^{\text{n}} \text{ of dissolved gas} = P_{\text{gas}} \times \text{solubility coeff}$$

This means that the higher is the partial pressure of gas adjacent to a liquid surface (e.g. in the lungs), the greater the concentration of gas that will dissolve (→ oxygen therapy!).

Concentrations of blood gases are usually stated as "partial pressures" rather than as "dissolved concentrations".

This means that if the partial pressure of O_2 in the alveolar air is 100 mm Hg (13 kPa) after inhalation, the arterial blood that leaves the lungs will contain dissolved O_2 with a concentration of ~100 mm Hg.

Partial pressure is another way of expressing solution concentration (for gases in a solution).

7 Pressure Gradient

If the pressure between two places is different, pressure *gradient* = pressure *difference* ÷ the distance between two places (unit = kPa/m).

Fluids (= liquids or gases) flow from places of high pressure to places of lower pressure. They flow down the pressure gradient. (Why do we press down hard with the hand during CPR?)

So blood flows from the LV to the arteries. Thus, air moves into and out of the lungs.

The *rate* of fluid flow (unit ml/s) is proportional to the pressure *gradient* ($\Delta P \div l$).

Molecules *dissolved* in a gas or liquid undergo *net diffusion* along their concentration gradient: e.g. if 100 mm Hg O_2 is dissolved in alveolar fluid and 40 mm Hg O_2 is dissolved in the venous blood of the alveolar capillary, diffusion occurs from the alveolar fluid to the venous blood.

8 Factors That Affect Volume Flow Rate

Volume flow rate (V) is the number of litres flowing past per minute (rather than the speed of flow). It is the same thing as **cardiac output**:

$$CO(\text{ml/min}) = HR(\text{bpm}) \times SV(\text{ml}).$$

1. **Pressure gradient** is the *difference* in pressure in the fluid at either end of the tube divided by the length of the tube (units kPa/m or mm Hg/m).
2. **Resistance to flow** through a tube depends on the length of the tube (l), the radius of the tube (R) and the internal friction between fluid particles (the fluid's *viscosity*, η "eta").

Poiseuille's law relates the pressure gradient ($\Delta P/l$) and resistance to flow to the volume flow rate V.

$$(CO =)\ V = \frac{\Delta P \times R^4}{l \times 2.54 \times \eta} \quad (\text{in m}^3/\text{s})$$

"The **volume flow rate**, V, equals the **pressure drop** ΔP, divided by the **resistance to flow** ($2.54\ \eta$ l) $\div R^4$."

$$V \alpha 1/l, \quad V \alpha R^4, \quad V \alpha 1/\eta \quad V \alpha \Delta P$$

Pressure Gradient in the Body (V α $\Delta P/l$)

- If the pressure gradient doubles, other things being equal, so does the flow rate (V α $\Delta P/l$).
- While lying horizontally, the pressure gradient is due to the pumping action of the heart.

(16 kPa (120 mm Hg) in the aorta to 4 kPa (30 mm Hg) at the start of the capillaries

2 kPa (15 mm Hg) at the end of the capillaries about 530 Pa (4 mm Hg) or less at the right atrium)

Radius of Tube (V α R^4)

- A wider tube produces less friction; in fact, if the radius is doubled (vasodilation), the flow rate increases 16 times! ($V \propto R^4$).

Conversely, if the radius is halved, the flow rate decreases to 1/16! (Coronary arteries narrowed by plaque lead to angina and myocardial infarction (MI).)

(vasoconstriction and vasodilation are potent mechanism to alter blood flow)

Blood viscosity (η) (V α $1/\eta$)

It is ≈ 0.004 Pa s for blood (0.001 Pa s for water) but changes with its speed. Blood vessels are not rigid tubes. Flow is pulsatile.

The higher the viscosity, the lower the flow rate ($V \propto 1/\eta$).

Friction increases with the percentage of cells in the blood (99% of cells are red blood cells (RBC)).

A "normal" man has about 42% of his blood volume taken up by cells (**haematocrit** of 42).

Haematocrit of 15 = anaemia.
Haematocrit of 70 = polycythemia.

(dehydration or hypothermia will cause blood friction to increase)

Length of Blood Vessels ($V \propto 1/l$)

There is greater resistance in the peripheral circulation than in the pulmonary circulation because of the greater length of the blood vessels.

9 Capillaries

The purpose of the circulatory system is to deliver blood to the capillaries.

The primary function of the capillaries is the exchange of gases, nutrients and waste products between body cells and the blood.

Average length = 1 mm
Average diameter = 8–10 μm (RBC slip-through in a single file)

O_2, CO_2, amino acids, glucose, lipids, wastes, drugs, electrolytes and hormones pass between the capillary blood and interstitial fluid by **diffusion** along their concentration gradient (from where they are in high concentration towards where they are in lower concentration).

10 Fluid Movement Through Capillaries

"Equation of continuity":

$$\text{Volume flow rate } (\text{V}) = \text{Cross} - \text{section } (\text{CS}) \text{ area} \times \text{speed.}$$

In the aorta, the cross-section area = 4 cm^2, and the speed of flow = 30 cm/s.
$\therefore V = 30 \times 4 = 120$ ml/s.

The volume flow rate must be the same in the capillaries, but the CS area = 4500 cm^2, so in the capillaries:

$$\text{Speed} = \text{V} \div \leq \text{CS area} = 120 \div 4500 = 0.03\,\text{cm/s}!\ (= 0.3\,\text{mm/s}).$$

This slow speed of blood flow allows exchange through the capillary wall to occur.

The capillary "bed" (extensive branching networks of capillaries) provide a large surface area for diffusion and filtration; ∴ the exchange of materials is rapid.

Movement Across the Capillary Wall

Blood enters the capillary bed from the terminal arteriole at ~35 mm Hg. Precapillary sphincters open to allow blood into the capillary.

More sphincters open as more blood flow is required (if closed, blood bypasses the bed via the "thoroughfare channel").

Blood exits the bed into the post-capillary venule at ~17 mm Hg.

In addition, bulk fluid flow (H_2O) through capillary walls occurs due to pressure differences between the inside and outside of the wall.

At the arteriole end:

HP (35 mm Hg) – OP (26 – 1) = 10 mm Hg
(hence, fluid moves OUT of the capillary).

At venule end:

HP (17 mm Hg) – OP (26 – 1) = –8 mm Hg
(hence, fluid moves INTO the capillary).

- Excess fluid returns to the blood via lymph vessels.

11 Type of Flow

Laminar (smooth, streamlined) flow is usual in long smooth vessels.

Turbulent flow is when fluid continually mixes in swirls and eddies. Turbulence adds to the resistance to blood flow by increasing friction. It will occur if blood velocity is high (e.g. 30 cm/s in the aorta) and when blood passes through a narrowing or constriction in a vessel (a stenosis).

Turbulent flow is "noisy".

- The opening and closing of the heart valves produce turbulent flow (heart valve sounds).
- Auscultatory blood pressure measurements "listen" for turbulent blood flow. The cuff causes turbulent flow in the squashed brachial artery (Korotkoff sounds).

12 Pressure Lecture Revision: Homework Exercise 14

1. Define pressure and explain how it is different from force.
2. Convert a blood pressure measurement of 130 mm Hg/80 mm Hg to units of kPa.
3. Use $P = F/A$ to determine the pressure exerted during CPR for the two cases below. In each case, the resuscitator can apply a force of 200 N.

    (a) If the whole hand (area = 140 cm^2 = 0.014 m^2) is used
    (b) If only the "heel" of the hand is used (area = 40 cm^2)

4. What is meant by positive pressure?
5. Write out Pascal's principle. Apply it to bed sore prevention or glaucoma or Queckenstedt's test or the knee or to a foetus surrounded by amniotic fluid.
6. Use Henry's law to describe why breathing air enriched to 30% O_2 is often a beneficial therapy.
7. Describe inhalation and exhalation using Boyle's law and pressure gradient.
8. In about four sentences, describe and explain the movement of fluid into and out of capillaries.

Lecture 22: Control of Blood Pressure

1 Introduction

Blood flows only when its pressure is higher at one place than in another, and it flows always from the place of higher pressure towards the place of lower pressure.

Blood pressure (BP) is highest in the arteries and falls progressively with distance from the heart. Arterial blood pressure is pulsatile, with waves of pressure being created by the contraction and relaxation of the ventricles.

The skeletal muscle pump and valves aid venous return in the limbs.

2 Arterial Blood Pressure (ABP)

ABP depends on:

(a) The compliance of the elastic arteries close to the heart (peripheral resistance (PR)).
(b) The volume of blood forced into them (cardiac output (CO)).

Blood contained in the aorta exerts a "backpressure" (about 80 mm Hg) on the aortic valve. This "afterload" must be overcome by the myocardium before blood can be ejected from the left ventricle (LV). Afterload results from the resistance to blood flow causing a limit to how fast the blood in aorta can drain away from heart.

LV begins to contract, the mitral valve closes and the myocardium undergoes **isovolumetric contraction** until ventricular pressure is greater than that of the afterload.

Arterial blood pressure is continuously oscillating between maximum and minimum values.

M. Caon, *Lecture Notes, Worksheets, and Exercises for Basic Anatomy and Physiology*, https://doi.org/10.1007/978-3-031-56296-9_15

Systolic pressure: the peak pressure in the blood due to the contraction of LV (~120 mm Hg while resting). It is actually 120 mm Hg *greater* than atmospheric pressure, denoted as +120 mm Hg.
Diastolic pressure: the minimum pressure in the aorta prior to LV contraction (~+80 mm Hg while resting).
Pulse pressure = the difference between systolic and diastolic pressure.
Mean arterial pressure (MAP): pressure remains closer to diastolic P than systolic P during the greater part of the cardiac cycle.

$$\text{MAP} = \text{diastolic pressure} + {}^{1}/_{3}\,\text{pulse pressure.}$$

Capillary pressure: ~40 mm Hg at the arterial end and ~20 mm Hg at the venous end.
Venous blood pressure: ~20 mm Hg in venules and ~0 mm Hg at vena cavae.

3 Hydrostatic Pressure

Pressure in the fluid increases with depth below the surface of the liquid.

The feet may be 120 cm below the level of the heart when standing. Consequently, BP in the veins and arteries of the feet may be at a pressure of +120 cm H_2O (= +90 mm Hg) simply due to the weight of blood in the vessels.

Hence, when one states that arterial pressure is 100 mm Hg, it usually means that this is the pressure at the "hydrostatic level" of the heart, i.e. as if supine.

4 Factors Influencing Blood Pressure

Three factors (and posture) can be identified as influencing BP in the closed cardiovascular system:

(i) **Cardiac output** (CO) – this is the amount of blood ejected each minute by the left ventricle into the aorta. It is calculated by multiplying the **stroke volume** (millilitres pumped by each beat) by the **heart rate** (HR) (no. of beats/min). An increase in CO will increase BP. The "average" resting output is ~5 l/min.

$$\mathbf{CO = SV \times HR}$$

(ii) **Peripheral resistance** – this refers to resistance to blood flow due to the friction between the blood and the vessel walls. It depends on the **viscosity** (internal friction) of the blood and the diameter and length of the vessels, particularly the arterioles.

The smaller the diameter of the vessel is, the more resistance it offers to blood flow. The greater is the length of the blood vessels, the greater is the resistance to flow.

Pulmonary resistance < peripheral resistance.

Greater peripheral resistance to blood flow will increase BP in the arteries.

$$\mathbf{MAP = CO \times TPR}$$

(iii) **Blood volume** – decreased fluid volume within a closed system decreases the pressure on the vessels within that system.

(**Exercise**: In heavy exercise, arterial pressure will rise by 30–40% and blood flow increases.)

5 Measuring Blood Pressure

It is the auscultatory method to measure pressure in the brachial artery. It is taken with the arm at the same level as the heart!

The pressure cuff (of the sphygmomanometer) squashes the brachial artery when inflated and prevents blood flow.

The pressure gauge reads air pressure in the cuff.

If cuff P > systolic P, the brachial artery is squashed closed.

If cuff P < systolic P but ≫ diastolic P, the artery opens partially for the short time that BP exceeds cuff pressure.

Blood flow through a partially squashed artery is turbulent and "noisy" → Korotkoff sounds.

Hypertension (resting BP > 140/90 mm Hg) – the heart pumps against greater resistance; if it works harder, the left ventricle enlarges.

There are many known causes (hormonal, renal, neurogenic), but 95% are still unknown aetiologically (origin). The devastating long-term effects are blindness, renal failure, myocardial infarction (heart attack), cerebrovascular accidents (strokes) and aneurysms.

The risk factors include heredity, obesity, smoking, stress and high salt intake. Women appear to suffer far fewer complications from persistent hypertension than men.

Hypotension – excessive decrease in BP:

1. Due to haemorrhage
2. Orthostatic (postural) blood pressure on standing up, may lead to syncope (fainting)

6 Cardiovascular Regulation

Arterial pressure is not controlled by a single pressure-regulating mechanism – several interrelated systems each perform a specific function.

The cardiovascular function is regulated by:

- Local factors
- Neural mechanisms
- Endocrine mechanisms

If the blood volume is constant, mean arterial pressure (blood pressure) is determined primarily by how much blood is pumped into the arteries (CO) and how much blood leaves the arteries – determined by the resistance of the arterioles to the blood flow.

Recall #1 … **MAP(BP) = CO × TPR**.

Vasoconstriction causes the total peripheral resistance (TPR) to increase.
☞ An increase in PR causes BP to increase.

Recall #2 … **CO = SV × HR**

Hence if heart rate increases, CO increases
☞ An increase in CO causes BP to increase.

7 Local Factors (Autoregulation of Blood Flow)

In normal resting conditions, CO is stable and PR is adjusted within individual tissues to control local blood flow. Autoregulation responds to *local factors* – independent of systemic factors (neural and hormonal).

Blood flow through an organ is regulated by adjusting the diameter of local arterioles and pre-capillary sphincters. Local vasodilators act at the tissue level.

Chemical changes such as (1) decreased concentration of O_2; (2) increased CO_2, H^+, metabolites and K^+; and (3) osmolarity and the release of eicosanoids, bradykinin and NO all cause arteriolar dilation.

8 Short-Term Control of Blood Pressure (Neural Regulation)

– Mechanisms respond quickly – *in seconds*

Cardiovascular Control Centre (CVC)

CVC (= cardiac centre + vasomotor centre) is an autonomic centre – a group of neurons in the medulla oblongata (of the brain stem) that regulate the heart rate, contractility and blood vessel diameter (by acting on smooth muscle).

CV centre receives input from:

- Higher brain centres (the hypothalamus and cerebrum)
- **Baro**receptors and
- Chemoreceptors

The output from the CV centre flows along sympathetic and parasympathetic (= vagus nerve) fibres.

(a) Sympathetic impulses along cardioaccelerator nerves (increase CO):

- *Increase* heart rate via the sinoatrial (SA) node
- Shorten conduction time via the atrioventricular (AV) node
- Increase the force of myocardial contraction

(b) Parasympathetic impulses along the **vagus** nerve (decrease CO):

- *Decrease* heart rate and
- Decrease the force of contraction

(c) Sympathetic impulses along vasomotor nerves to blood vessel walls maintain moderate vasoconstriction (vasomotor tone):

- An *increase* in impulses causes generalised vasoconstriction (i.e. increased PR).
- A *decrease* in activity causes the dilation of arterioles, allowing faster blood flow out of arteries (i.e. decreased PR).

Baroreceptor Reflexes

Baroreceptors are pressure-sensitive neurons that monitor the stretching of blood vessel walls and atria. They are located in the wall of the carotid sinus, the aortic arch and the wall of most large arteries in the neck and thorax.

(a) Aortic reflex maintains general systemic BP, initiated by baroreceptors in the wall of the arch of the aorta.
(b) Carotid sinus reflex maintains normal BP in the brain.

(1) A rise in arterial pressure stretches baroreceptors. (2) They send a faster stream of impulses to the vasomotor centre (VM), (3) which inhibits it. (4) Hence, the VM centre sends fewer impulses to the vasculature, (5) so vasodilation results. (6) The decrease in PR causes a decrease in arterial pressure.

If BP falls, baroreceptor reflexes increase heart rate and the force of contraction and promote vasoconstriction.

Immediately upon standing, arterial pressure in the head and upper body falls (as blood pools in the lower extremities). Baroreceptors respond immediately (by decreasing their firing rate) and cause the CV centre to increase sympathetic discharge → HR increases, and vasoconstriction occurs (and decreases parasympathetic discharge), which minimises the decrease in pressure.

However, over 1–2 days of abnormal BP, baroreceptors are "reset" and cease to send impulses even if arterial pressure remains above 200 mm Hg. Hence, the prolonged regulation of arterial pressure needs other (renal) mechanisms.

An aside: During extended bed rest, the blood is distributed more evenly throughout the body than is the case when standing – gravity does not cause pooling in legs. This raises arterial pressure. Over ~3 days, the kidneys excrete what they perceive to be excess fluid, so blood volume decreases (by ~12%). When the person finally gets out of bed, due to decreased blood volume, baroreceptors cannot compensate for the orthostatic hypotension. The patient may feel dizzy/faint.

Chemoreceptor Reflexes

Chemoreceptors are located in two carotid bodies and several aortic bodies, close to baroreceptors.

They respond to the decrease of O_2, increase of CO_2 and decreased pH in the blood that occurs as arterial pressure falls. They transmit signals to the vasomotor centre, which *excites* it to produce vasoconstriction, and to the cardioaccelerator centre, which increases CO → more blood to the lungs.

An aside: Increased CO_2 in cerebrospinal fluid (CSF) also stimulates chemoreceptors in the medulla oblongata. This increases respiratory rate so **that increased CO is accompanied by an increased rate of resp**. so that CO_2 in the blood is blown off and oxygenation occurs.

9 Long-Term Control of Blood Pressure

Blood pressure can be stabilised or maintained within normal limits only when blood volume is stable.

Four hormones (angiotensin II, antidiuretic hormone (ADH), aldosterone and atrial natriuretic peptide (ANP)) regulate blood volume (and hence BP).

Antidiuretic Hormone (or Vasopressin)

ADH is released from the posterior pituitary in response to:

- A decrease in blood volume
- Increased blood osmolarity (>280 mosmol/l)
- Circulating angiotensin II

It causes an increase in permeability to the water of the collecting duct of the nephron → less water excreted in urine → blood volume does not decrease.

It also causes peripheral vasoconstriction and hence an increase in BP (an immediate response).

Angiotensin II
Angiotensin II appears in the blood after the protein enzyme **renin** is released by (the juxtaglomerular cells of) the kidney when arterial pressure falls.

Renin catalyses the formation of angiotensin I (from angiotensinogen in the blood), which is converted to **angiotensin II** (in small vessels of the lung) by an angiotensin-converting enzyme (**ACE**).

(What effect would the drug *Captopril* – an "ACE inhibitor" – have on blood pressure?)

Angiotensin II persists in the blood for ~2 min. It causes four things to happen:

1. The adrenal glands release **aldosterone**, which increases the re-absorption of Na^+ from the filtrate by kidney tubules.
2. ADH is secreted by the posterior pituitary (hence kidney tubules become permeable to water, which osmotically follows Na^+ out of the filtrate).

   Thus, urine production decreases as water is reclaimed from the filtrate and returns to the blood – this maintains the circulating blood volume.
3. In addition, angiotensin II stimulates thirst → we drink, which increases extracellular volume and hence blood volume.
4. Angiotensin II is also a vasoconstrictor. It causes:
   - The rapid, intense vasoconstriction of arterioles (increases PR).
   - Mild vasoconstriction in the veins (promotes venous return).
   - *Constriction of renal blood vessels, decreasing blood flow to the kidneys, hence decreasing the glomerular filtration rate*. This decreases the excretion of salt and water.

Atrial Natriuretic Peptide (ANP)
(Produced by muscle cells in the right atrium in response to excessive stretching of atria)

It acts to **reduce** blood volume and blood pressure by:

- <u>Increasing Na+ excretion in the kidneys</u>
- Increasing the volume of urine produced (↑GFR)
- Reducing thirst
- Blocking the release of ADH, aldosterone, adrenaline and noradrenaline (vasoconstrictors)
- Stimulating peripheral vasodilation.

10 Level of Water and Salt Intake

If the kidney is normally healthy, then the level of water and salt intake determines the long-term arterial pressure.

(If well hydrated) Water is normally excreted by the kidneys as rapidly as it is ingested.

Salt is not so readily excreted in the following:

- If excess salt is present in the diet, the osmolarity of the extracellular body fluids (blood) increases. Hence, water moves out of the cells into the extracellular fluid (ECF) → blood volume increases.
- Osmolarity that is >280 mosmol/l stimulates the thirst centre, and you drink more in order to reduce the increasing osmolarity. This water is absorbed into the blood.
- Hence, extracellular fluid volume increases (so blood volume increases).
- Concurrently, the increased osmolarity stimulates the pituitary to secrete more ADH.
- ADH causes the kidneys to re-absorb more water before it is excreted as urine.
- Hence, the extracellular fluid *volume* increases (so blood volume increases).

(Note: Drinking – not decreased osmolarity – relieves thirst. Receptors in the mouth respond to cold water by decreasing thirst. Hence, sucking ice chips relieves thirst without putting significant amounts of fluid in the body.)

This salt-driven increase in blood volume means that venous return must increase.

Hence, cardiac output increases, which causes arterial pressure to increase.

(Note: The advice to decrease salt intake to avoid high blood pressure is for this reason.)

11 Hypertension

MAP >110 mm Hg when resting, or
Systolic pressure >140 mm Hg, and
Diastolic pressure >90 mm Hg.

The moderate elevation of arterial pressure leads to shortened life expectancy. If MAP is 50% or more above normal, a person can expect to live for only a few more years. Death occurs due to either of the following:

1. Excess work on the heart leads to LV hypertrophy (more muscle → more O_2 needed, coronary artery cannot supply → ischemia). If PR remains high, the heart is unable to cope with the workload. LV pumps less blood than RV! This creates pulmonary oedema, then the development of congestive heart failure, coronary heart disease or both (→ heart attack).
2. The high pressure ruptures a major blood vessel in the brain (→ cerebral infarct).
3. Very high pressure causes multiple haemorrhages in the kidney (→ kidney failure).
4. Hypertension in arteries damages the endothelial lining of vessels → atherosclerotic plaque.

Drugs to Treat Hypertension

"Essential hypertension" (hypertension of unknown origin) may be treated by drugs that reduce the re-absorption of salt and water in the kidneys, called **diuretics** (or natriuretics: natrium = Na = sodium).

Calcium Channel Blockers
These drugs can be used to treat hypertension. They block the entry of Ca^{++} ions to the vascular smooth muscle (recall that Ca^{++} ions are necessary for muscle contraction) so it cannot contract → arterioles dilate, causing peripheral resistance to decrease.

Beta Blockers
The neurotransmitters adrenalin and noradrenalin attach to the beta-1 receptor sites of the heart and cause HR and contractility to increase. Drugs called "beta blockers" may be used to treat hypertension. These drugs attach to the beta-1 receptors; hence, they block adrenalin's access to the receptors. Thus, heart rate and arterial pressure are not increased.

Mineralocorticoid receptor antagonists (MRAs) block the action of aldosterone by preventing its binding to the receptor.

The aldosterone synthase inhibitor prevents the synthesis of aldosterone.

ACE Inhibitors

e.g. captopril

Angiotensin II Antagonists
Also called Angiotensin receptor blocker (ARBs, molecules that block/prevent AII binding to the receptor and hence stimulating it – antagonists(=blocker) do not stimulate the receptor), e.g. *irbesartan.*

Which "blood pressure" do you mean?

- Mean arterial pressure
- Systolic arterial pressure (while supine)
- Systolic arterial pressure (while seated)
- Diastolic arterial pressure (while supine)
- Diastolic arterial pressure (while seated)
- Blood pressure during exercise!
- Capillary blood pressure
- Central venous pressure

(From haemodynamic monitoring)

- Right atrial pressure (2–8 mm Hg)
- Right ventricular systolic pressure (20–30 mm Hg)
- Right ventricular diastolic pressure (8–15 mm Hg)
- Pulmonary artery systolic pressure (20–30 mm Hg)
- Pulmonary artery diastolic pressure (8–15 mm Hg)

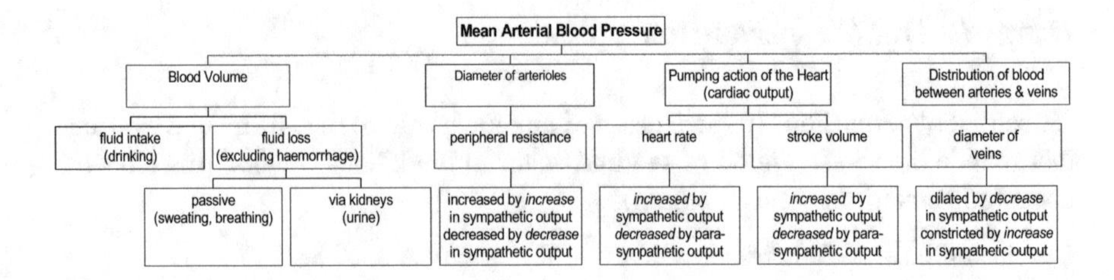

Factors affecting mean arterial pressure

12 Control of Blood Pressure Revision: Homework Exercise 15

1. Define mean arterial blood pressure.
2. What are the three basic factors that determine arterial blood pressure?
3. State the formula that relates peripheral resistance, stroke volume and heart rate to mean arterial blood pressure.
4. What effect does vasoconstriction have on arterial blood pressure and why?
5. What are the effects on blood volume and arterial pressure of the following:

   (a) Increased GFR?
   (b) Decreased GFR?
   (c) An increase in sympathetic stimulation to the heart and vascular smooth muscle?
   (d) An increase in parasympathetic stimulation to the heart?
   (e) ADH secretion?
   (f) Drinking water?

6. (a) What is the stimulus for renin release?
   (b) How does angiotensin II arise from the release of renin?
   (c) What effects do angiotensin II have?
   (d) How do the drugs known as ACE inhibitors work?
   (e) Why do "calcium channel blockers" help reduce blood pressure?
7. Name the major mechanisms for the short-term regulation of blood pressure.
8. Write a paragraph about the role of the kidneys in the long-term control of blood pressure.

Lectures 23 and 24: Anatomy and Physiology of Respiration

Functions of the Respiratory System

1. Exchange of gases between the atmosphere and the blood in alveolar capillaries.
2. Regulation of blood pH.
3. Protection from inhaled pathogens and irritants (+ sneezing and coughing).
4. Vocalisation and smell!
5. Alveolar capillary endothelial cells produce angiotensin-converting enzymes.

1 Respiration

(Means many things)

- Ventilation of the lungs (breathing).
- External respiration: exchange of gases in the lungs (between alveolar air and blood in pulmonary capillaries).
- Internal respiration: exchange of dissolved gases in body tissues (between capillary blood and body tissues).
- Cells produce adenosine triphosphate (ATP) (carbon dioxide (CO_2)) from organic molecules by using oxygen (O_2) (cellular respiration).

The goal of respiration is to control the concentration of O_2, CO_2 and H^+ ions dissolved in the blood. Respiratory activity responds to changes in these.

M. Caon, *Lecture Notes, Worksheets, and Exercises for Basic Anatomy and Physiology*, https://doi.org/10.1007/978-3-031-56296-9_16

2 Anatomy

Upper respiratory tract (= nose, pharynx, larynx)
Lower respiratory tract (= trachea, bronchi, bronchioles, alveoli)
Pleurae (left lung 87% size of the right lung (RL))

Nose (is between the nares and choanae): It consists of: nares (nostrils), external nose (composed of cartilage), nasal cavity contains threefold of tissue (conchae) through which pass three passages (meatus) for air. Air passing through a meatus contacts the mucus membrane and is warmed, humidified and filtered.

The olfactory mucosa (smell) on the cribriform plate of ethmoid bone is a route for infection.

Extensive plexus of thin-walled veins under the epithelium ⇒ nose bleeds easily.

Sinuses mucus-membrane-lined cavities in the skull bones – frontal, ethmoid, sphenoid and maxillary; drain into the nose via small holes

Pharynx (between the choanae and larynx): nasopharynx connected to the middle ear by the Eustachian tube (auditory tube)

Larynx (voice box): a cartilaginous chamber; upper opening = glottis, lower opening = trachea.

The epiglottis covers the glottis during swallowing. It functions to ensure that only air enters the trachea and to keep the airway patent. It also produces sound (our voice)

Vestibular folds attached to the vocal cords also close off the glottis.

The laryngeal prominence of the thyroid cartilage = "Adam's apple".

The lower respiratory tract has a **conducting zone** (moves gases to and from the respiratory zone) and a **respiratory zone** (alveoli where gas exchange takes place).

Trachea (windpipe): kept open by "C" shaped rings of the cartilage. The gap in "C" closed by a muscle allows the oesophagus to expand as food bolus passes.

L and R – **primary bronchi**
3R and 2L – **secondary bronchi** (one per lobe)
10R and 8L – **tertiary bronchi**

The bronchi is supported by cartilage. The trachea and bronchi have muscles between the cartilage.

Bronchioles are <1 mm diameter and have smooth muscle but NO cartilage in walls.

A cross-sectional area for airflow is the greatest in bronchioles and least in trachea!

A bronchiole branches into 50–80 **terminal bronchioles** (lobule = the smallest section of the lungs, supplied by a terminal bronchiole).

Terminal bronchioles divide into **respiratory bronchioles** with little smooth muscle (then into *alveolar ducts*, which end in *alveolar sacs* – *alveoli* open from these sacs).

There are ~65,000 terminal bronchioles. The contraction of their smooth muscle causes many obstructive lung diseases.

There are 20–25 "generations" of branching before air reaches alveoli.

There are about 3×10^8 alveoli in the two lungs. An alveolus is a pouch ~0.2–0.5 mm in diameter, formed by the respiratory membrane.

Mucus (secreted by goblet cells) coats all passages from the nose to terminal bronchioles. The bronchial tree is lined with **ciliated cells** except for most distal respiratory bronchioles. The cilia move mucus and trapped particles up the bronchial tree and keeps airways moist.

(Cilia are very important in moving mucus out of the bronchial tree – the accumulation of mucus leads to persistent cough and recurrent infection (bronchiectasis); see cystic fibrosis.)

Lungs and Pleura

The lungs extend from the apex (superior to the clavicle) to the base at the sixth rib (medially) and to the seventh rib laterally. The left lung is smaller than the right.

The **hilum** (hilus) is the point where the primary bronchus and pulmonary artery/veins enter and leave the lungs. The lungs are suspended from the hilum and adhere to the chest walls via pleurae. A pleura is a (two-layer) serous membrane (a sac) that surrounds each lung separately.

The *visceral* pleura adheres to the lung, while the *parietal* pleura adheres to the chest wall. In between is a small amount (a few millilitre) of slippery pleural fluid – lubricates movement between the chest and the lung. It also ensures that when the chest expands, so do the lungs (a negative pressure is maintained between the pleurae).

If air enters between two pleurae (e.g. from a wound) → **pneumothorax** and the lung collapses.

3 Respiratory Membrane

The total surface area for gas exchange >70 m^2. Alveolar surfaces enclose about 2700 ml of alveolar air "in contact" with 60–140 ml blood.

The respiratory membrane is only 0.5 μm thick and consists of fluid and surfactants, epithelial cells of alveolus, fused basement membranes (of the alveolar epithelial cell and the capillary endothelial cell) and capillary endothelial cells. A small distance allows the rapid diffusion of O_2 and CO_2 between the capillary blood and gas in an alveolus.

Red blood cells (RBC) remain in alveolar capillaries for ~0.8 s (less if cardiac output is high), O_2 and CO_2 exchange takes about 0.3 s.

Alveolar epithelial cells produce an *angiotensin-converting enzyme*, while some (septal cells) produce a "**surfactant**", i.e. a detergent that reduces the surface tension of alveolar fluid – allows the alveolus to expand more easily against the tendency of surface tension to contract alveoli. Its absence in premature babies →"respiratory distress syndrome".

Alveolar **macrophages** patrol inside of the alveolus engulfing inhaled infectious organisms. They are swept up the bronchial tree by cilia.

Gas Exchange Through the Respiratory Membrane

Standard atmospheric pressure is 1016 hPa (760 mm Hg), of which 79% (600 mm Hg) is due to N_2 and 21% (160 mm Hg) is due to O_2.

In alveoli, gas "partial pressures" are $N_2 = 570$, $O_2 = 104$, $CO_2 = 40$ and $H_2O = 47$ mm Hg (570 + 104 + 40 + 47 = 760 mm Hg)

Note well!

1. Gases (even when dissolved in liquid) move from where they are at high partial pressure to where they are at low partial pressure. For example, *if gas in alveoli has $pO_2 = 104$ and blood contains O_2 at the concentration of 40 mmHg, then O_2 will move into the blood from the alveolar gas –* ***and then from the plasma onto haemoglobin (Hb)***.
2. Gases dissolve in liquid in proportion to their partial pressure in the gas mixture (Henry's law).

Henry's law: Amount of diss $O_2 = k \times pO_2$

For example, if blood entering the lung has $pCO_2 = 46$ and alveolar air has pCO_2 at 40 mmHg, then CO_2 will move out of the blood into the alveolar gas until the CO2 pp in gas and in the blood are the same.

In **de-**oxgenated blood **entering** the lungs:

$pO_2 = 40$ mm Hg, $pCO_2 = 46$ mm Hg
(∴ Hb 75% saturated)

Air in alveoli:

$pO_2 = 104$ mm Hg, $pCO_2 = 40$ mm Hg

In oxygenated blood **leaving** the lungs:

$pO_2 = 100$ mm Hg, $pCO_2 = 40$ mm Hg
(∴ Hb 99% saturated)

Gas exchange through the respiratory membrane is efficient because (/if!):

1. Gas pressure gradients (ΔpO_2) are substantial.
2. Diffusion distance is small (x).
3. O_2 and CO_2 are lipid soluble.
4. The total surface area of alveoli (A) is large.
5. Blood flow and airflow are coordinated.

Fick's law (for rate Q, of diffusion):

$$Q \quad \frac{pO_2 \quad A \quad d}{x} \qquad d \text{ is diffusion coeff.}$$

4 Pulmonary Ventilation

We exhale (some) alveolar air – high in CO_2 and deficient in O_2 – and inhale some fresh air that mixes with residual alveolar volume.

Normal quiet breathing: 12–18 bpm and ~500 ml/breath; **diaphragm** is moved down, **external intercostals** are contracted to move the sternum forward and the ribs outward. This increases the volume of the chest; hence, alveolar air pressure decreases (Boyle's law) by 400 Pa (3 mm Hg).

Boyle's law: $P \propto \frac{1}{V}$

This pressure difference forces air into the lungs. Normal exhalation is accomplished passively by the elastic recoil of the thorax.

Deep inspiration is aided by neck muscles (*scalene*, *sternocleidomastoid*) and the *pectoralis minor* and *serratus anterior*.

Forced exhalation uses *internal intercostals*, *external and internal obliques*, and *rectus* and *transversus abdominus*.

Innervation

The diaphragm is innervated by **phrenic** nerves arising from C3–C5 vertebrae! Intercostals are innervated by spinal nerves that arise from each thoracic vertebrae.

Airway Resistance

The change in lung volume for a given pressure difference (between the atmosphere and lungs) is called **compliance** (= ease with which the lung can be expanded). Low compliance means it is difficult to inflate the lungs ⇒ poor ventilation.

In healthy people, resistance to airflow is mainly in large airways (bronchi) as their X-sectional area is < that of the 65,000 bronchioles.

In diseases, small bronchioles are easily occluded, contributing to airway resistance. The smooth muscle of bronchioles constricts them easily (no cartilage to hold them open!).

Bronchoconstriction is triggered by parasympathetic nerve impulses, histamine, smoke and dust (airborne irritants).

Bronchodilation (↓ resistance) is stimulated by sympathetic nerve impulses, adrenalin and noradrenalin.

Alveolar Ventilation

Air that gets into the alveoli is available for gas exchange.

Air in the conducting zone of the bronchial tree is (anatomical) **dead space** – this air cannot exchange gases.

Physiological dead space = anatomical dead space + pathological dead space due to non-functioning alveoli.

About 150 ml of the 500 ml of resting inhalation is "dead air". Only ~13–15% of lung air is changed over on one breath at rest.

Deep breathing (>500 ml) increases the *proportion* of fresh air to dead air that enters alveoli – rapid shallow breathing (<500 ml) increases the *proportion* of already-breathed air in the dead space that re-enters alveoli to fresh air.

5 Regulation of Respiration

The nervous system automatically adjusts the rate of ventilation so that blood O_2 and CO_2 concentration is maintained within homeostatic limits even during heavy exercise.

Normal *arterial* blood gas values:

CO_2 = 35–45 mm Hg
O_2 = 90–110 mm Hg

"**Respiratory centres**" of the brain are located in the brain stem (medulla oblongata and pons). They respond to incoming signals from *peripheral* chemical receptors (when $CO_2 \uparrow$, pH $\downarrow$ or $O_2 \downarrow$) by sending signals to respiratory muscles via nerves.

H^+ (and CO_2) affect the respiratory centre directly. H^+ cannot cross the blood-brain barrier, but CO_2 does. CO_2 in cerebrospinal fluid (CSF) produces H^+ – *central* chemoreceptors are very responsive to H^+ as there is no buffer in CSF. Thus CO_2 (via H^+) is the main controller of ventilation.

Anaesthesia or narcotics (e.g. morphine) can depress respiratory centres, so can they raise intracranial pressure or brain oedema.

If arterial blood O_2 concentration falls to <60 mm Hg (a rare occurrence), peripheral chemoreceptors (in aortic bodies and carotid bodies) are strongly stimulated → impulses to respiratory centre.

Eupnoea = normal resting breathing (12–20 bpm)
Bradypnoea = abnormally slow (<12 bpm)
Tachypnoea = abnormally fast (>20 bpm)
Dyspnoea = laboured, difficult breathing
Apnoea = cessation of breathing!

6 Gas Transport in Blood

N_2 is inert in the body.

CO_2 is much more soluble than O_2.

Transport of Oxygen in the Blood
97% of O_2 is carried bound to haemoglobin (Hb) in RBC, only 3% dissolved in plasma.
Oxygen binds loosely and reversibly to Hb:

when blood pO_2 is high (in alveolar capillaries), O_2 binds (or stays bound) to Hb;
when blood pO_2 is low (in systemic capillaries), O_2 is released from Hb.

Percentage <u>saturation</u> of Hb with bound O_2:

95–99% in the arterial blood
75% for venous blood (venous reserve!)

Normal Hb concentrations in RBC:

♂ = 13–17.5 g/ml, ♀ = 12–16 g/ml

Transport of Carbon Dioxide in the Blood CO_2 leaves the cell as dissolved gas.

Seven per cent (7%) are transported in solution in plasma.
Twenty-three per cent (23%) are bound to haemoglobin ($HbCO_2$) in RBC.
Seventy per cent (70%) reacts with water to form carbonic acid.
In RBC, "carbonic anhydrase" (enzyme) increases the rate of formation of H_2CO_3 5000-fold!

Carbonic anhydrase

$$CO_2 + H_2O \rightarrow H_2CO_3 \rightarrow HCO_3^- + H^+$$

H_2CO_3 then dissociates into bicarbonate (HCO_3^-) and H^+.
<u>HCO3− moves out of RBC into plasma</u> in exchange for Cl^- (the "chloride shift" maintains electrical neutrality).
<u>H+ stays in RBC and binds with Hb</u> after it releases O_2 to the tissues (i.e. H^+ is buffered by Hb and does not contribute to the acidity of tissues).
In alveoli, O_2 binds to Hb and reverses the above reactions.

7 Spirometry

Respiratory volumes are measured using a spirometer:

<u>Tidal volume</u>: ~500 ml (volume inhaled/exhaled in one quiet breath)
<u>Inspiratory reserve volume</u>: ~3 L (volume in excess of tidal that can be inhaled with max effort).
<u>Expiratory reserve volume</u>: ~1.2 L (volume in excess of tidal that can be exhaled with max effort).
<u>Residual volume</u>: ~1.2 L (volume of air remaining in the lungs after max effort exhalation) (difficult to measure)

Vital capacity ERV TV IRV.

Total lung capacity RV VC 5.9L male,4.3L female .

Flow rates, e.g. one of which is: FEV_1 = maximum volume of air that can be forcefully exhaled in 1 s.

(Used to measure the degree of pulmonary obstruction, e.g. with asthma. Healthy lungs can expel ≥80% of their volume in 1 s.)

Forced vital capacity (FVC) is decreased in obstructive lung disease (forced expiratory flow at 25%, 50% and 75% of FVC reflects the obstruction of large, medium and small airways, respectively). *Calculating the ratio between forced expiratory volume in the first second (FEV1) and FVC is particularly helpful in deciding whether a person's lung disease is an obstructive or restrictive type.*

(Obstructive lung diseases include conditions (e.g. asthma, bronchitis, bronchiolitis) that make it hard to exhale all the air in the lungs. People with restrictive lung disease (e.g. pulmonary fibrosis, asbestosis) have difficulty in fully expanding their lungs with air.)

8 Additional Information

Lung cancer is the leading cause of cancer (ca) mortality in men and women **in the USA** (30% of all ca deaths).

Eight per cent (80%) is non-small cell lung cancer (NSCLC) (no curative therapy for locally advanced stage); 20% is small-cell lung ca

More women die from lung ca than from breast, ovarian and uterine ca **combined**.

Four times as many men die from lung ca than from prostate ca. (Do not smoke.)

Lung cancer has been the most common cancer in the world for several decades, and by 2008, Lung ca were 12.7% of all new cancers. 18.2% of all cancer deaths were due to lung cancer.

NSCLC represents approximately 80% to 85% of all lung cancers. Unfortunately, at the time of diagnosis, approximately 70% of NSCLC patients already have advanced or metastatic disease not amenable to surgical resection.

(NSCLC is the most common type of lung cancer, and it accounts for 85% of lung cancer cases in the United States.)

The mean age at diagnosis is 61.6.

The median overall survival (OS) (95% confidence interval) from the first line of therapy (N = 694) was estimated at 20.27 (18.27, 22.70) months.

Most of the NSCLC patients were former (n/N = 382/758, 50.4%) or current smokers (n/N = 216/758, 28.5%).

Chronic obstructive pulmonary disease (COPD) is the fourth leading cause of morbidity and mortality in the USA. It affects ~20% of chronic smokers. It involves the airways of all size, ultimately leading to alveolar destruction and the loss of gas-exchange capacity, dyspnoea, limited exercise tolerance, mucus hypersecretion, cough and poor quality of life.

Pulmonary circulation contains ~0.5 L of blood (10% of total), with ~75 ml in capillaries spread out over ~60 m^2 contact with alveoli.

Pulmonary circulation flows very rapidly.

Pulmonary blood press is ~ 25/8 mm Hg.

As much blood flows through the lungs as through the rest of the body, that is, 5 L/min.

As a breath is inhaled, atmospheric air mixes with (1) water vapour that evaporates from the epithelium of the nasal passages, (2) CO_2 that enters the alveoli from lung capillaries and (3) the residual gas volume of the lungs from which O_2 has left the alveoli for the capillaries.

These gases dilute the inhaled air and contribute to the pressure of gas in the alveoli so that the pO_2 in the alveoli is less than that in the atmosphere.

Chronic obstructive pulmonary disease (COPD), a common preventable and treatable disease, is characterized by persistent airflow limitation that is usually progressive and associated with an enhanced chronic inflammatory response in the airways and the lung to noxious particles or gases.

Cigarette smoking is the most common risk factor for COPD. Exacerbation and comorbidities contribute to the overall severity in individual patients. COPD is a major cause of morbidity and mortality worldwide and results in economic and social burden, which is both substantial and increasing.

COPD is characterized by structural changes in the airways resulting from repeated injury and repair and by bronchoconstriction, which is an important target for pharmacologic interventions.

Dyspnoea, chronic cough and sputum production are the most common clinical symptoms.

Transport of CO_2 from Tissues to Lungs in the Blood

- Cells produce CO_2, which dissolves in water and diffuses through the plasma membrane out of cells.
- Some CO_2 dissolved in blood is carried as dissolved gas to the lungs (7%).
- CO_2 dissolved in plasma reacts with H_2O to form H_2CO_3 (slow reaction).
- Dissolved CO_2 diffuses into RBC, where *carbonic anhydrase* converts CO_2 to H_2CO_3 very rapidly (70%).
- Also, in RBC, H_2CO_3 dissociates into HCO_3^- and H^+.
- Most of this H^+ combines with haemoglobin (it is a protein buffer).
- HCO_3^- diffuses out of RBC into plasma (while Cl^- moves into RBC, thanks to a bicarbonate-chloride carrier protein known as the "chloride shift").
- This allows *more* H_2CO_3 in RBC to dissociate into HCO_3^- and H^+.
- In addition, CO_2 combines (in a slow reaction) **directly** with haemoglobin to form a loosely bound carbaminohaemoglobin compound. Furthermore, a small amount of CO_2 combines **directly** with plasma proteins (15–25%).

The Carbon Dioxide (Carbonic Acid)/Bicarbonate Buffer System

Carbonic acid is a weak acid (it can destroy a base and in the process transform into a bicarbonate ion), and bicarbonate ion is a weak base (it can destroy acid and in the process transform into a carbonic acid molecule). Carbonic acid is unstable, so it "exists" as $CO_{2(aq)}$ in the blood, and because of the work of **carbonic anhydrase**, we can use $CO_{2(aq)}$ as a stand in for H_2CO_3.

$CO_{2(aq)}$ is able to neutralise hydroxide *directly* by the reaction:

$$CO_{2(aq)} + OH^- \leftrightarrow HCO_3^{-}{}_{(aq)}$$

H^+ appears because we eat much acidic food, and many products of metabolism are acidic.

HCO_3^- appears because our body makes it in kidney tubules, in cells lining the gut, in the pancreas, in the liver (bile) ?? and in RBC.

CO_2 appears since it is the product of using oxygen in cells.

Once in existence, CO_2, HCO_3^- and H^+ can participate in the reversible reactions below:

$$\begin{array}{l} CO_2 + H_2O \leftrightarrow H_2CO_3 \leftrightarrow HCO_3^- + H^+ \\ 400 \qquad\qquad\quad 1 \qquad 8000 \end{array}$$

Of the CO_2 that is dissolved in plasma, some reacts with H_2O to form H_2CO_3 (slow reaction) some of which in turn can dissociate into HCO_3^- + H^+. (Carbonic acid is a **very weak acid,** so very little of it dissociates into ions!)

However, H_2CO_3 also is unstable, so it dissociates rapidly into CO_2 and H_2O (ratio of H_2CO_3:CO_2 = 1:400). The net result is ratio of CO_2:HCO_3^- = 1:20.

As CO_2 concentration increases, more H_2CO_3 is produced (particularly in RBC) and consequently more HCO_3^- and H^+ too. This would decrease blood pH EXCEPT for reaction in RBC and for protein and phosphate buffers.

When acid is neutralised by the bicarbonate buffer, the blood's $CO_{2(aq)}$ concentration increases.

$CO_{2(aq)}$ can "disappear" by being breathed out as gas, and H^+ can "disappear" by being excreted in urine.

Laplace's law for cylindrical membrane (capillary): membrane tension, $T = P \times r$.

Note 1: Capillaries have a small r, so tension in the capillary wall is smaller for a given pressure than in a more thick-walled larger diameter artery.

Note 2: An aneurysm has a large r with the same pressure as in the rest of the artery, so the wall is at greater tension and greater risk of bursting.

Laplace's law for spherical membrane with one surface (alveolus): membrane tension, $T = \frac{P \times r}{2}$.

Note 1: The surface tension of the liquid provides the necessary membrane strength to maintain the alveolus.

Note 2: For two connected "spheres" containing gas, pressure in both is the same (Pascal's principle) – changes in wall tension do not arise from differences in pressure.

Note 3: If surface tension is too high (insufficient surfactant), the radius of alveoli will decrease for the same pressure (alveoli collapse). A premature baby has insufficient strength to expand the lungs.

Note 4: If surface tension is too low (excess surfactant), the radius of the alveoli will increase for a given air pressure.
Note 5: Bubbles are unstable, and the surfactant/alveolar radius self-regulates – as the alveoli radius decreases, the surfactant gets more concentrated, lowering the surface tension and halting its collapse (and vice versa).

COPD results from inflammatory damage of the airways and alveoli after chronic exposure to noxious airborne particles, principally from cigarette smoking. It comprises several conditions:

- **Chronic bronchitis** – persistent and recurring inflammation of the bronchi with the consequent release of active enzymes to the surrounding tissues
- **Emphysema** – abnormal and permanent enlargement of the airspaces distal to the terminal bronchioles with destruction of their walls
- **Peripheral airway disease** – histopathologically characterized by an increase in goblet cells, intraluminal mucus, inflammation, increased smooth muscle mass, fibrosis resulting in the narrowing and obliteration of small airways.

9 Respiratory System Revision: Homework Exercise 16

1. Define the following: dead space, lung compliance and respiratory distress syndrome.
2. How does the construction of the walls of bronchi and bronchioles differ?
3. What can cause bronchodilation, and what can cause bronchoconstriction?
4. Why does hyperventilation (= rapid shallow breaths) result in an increase in dissolved CO_2 in the blood?
5. To what do the chemoreceptors in the respiratory centre of CNS respond? (And explain your answer.)
6. How and why is the composition of alveolar air different from atmospheric air?
7. What is FEV1, and why is it decreased in obstructive diseases such as asthma?
8. Describe the chemical changes that occur in the RBC that facilitate carbon dioxide transport.
9. State Henry's law. Use Henry's law to describe why breathing air with 30% O_2 is often a beneficial therapy.
10. State Boyle's law. Describe inhalation and exhalation using Boyle's law and pressure gradient.
11. What are the functions of the mucous glands and the ciliated epithelial cells?
12. Give definitions for the following: eupnoea, apnoea, bradypnoea, tachypnoea and dyspnoea.
13. Why would you encourage an anxious person to breathe deeply rather than with shallow breaths? (Hint: think about "dead-space".)
14. What would you anticipate would be the effect of pneumothorax (air in the intrapleural space) on the lungs and breathing?
15. What could you anticipate would be the effect on breathing of a spinal cord injury at the level of the sixth cervical vertebra? (Hint: from where is the diaphragm innervated?)

Lectures 25 and 26: Reproductive System

(i) The purpose of sexual reproduction is to produce offspring that are genetically different from each other and from their parents.
(ii) Puberty confers the ability to produce gametes capable of being fertilised.

1 Male Anatomy

- It consists of two testes, two epididymis, a scrotum, two vas deferens, two seminal vesicles, two ejaculatory ducts (prostate glands), urethra, two bulbourethral glands, a penis and an external urethral meatus.

2 Male Reproductive Tract

Seminiferous tubules produce sperm, and cilia produce currents that transport them to the epididymis.

The **epididymis** stores sperm, regulates their fluid environment and facilitates their maturation.

The **vas (ductus) deferens** stores sperm and transports the sperm by peristaltic contractions.

(The inguinal canal lies between the deep inguinal ring and the superficial inguinal ring. In men, it contains the spermatic cord (= vas deferens, blood vessels, lymphatics, nerves). In women, it contains the round ligament of the uterus.)

The **ejaculatory ducts** (~2 cm long) begin at the end of the vas deferens. Here, seminal vesicles empty seminal fluid into the tract, and it mixes with and dilutes sperm.

M. Caon, *Lecture Notes, Worksheets, and Exercises for Basic Anatomy and Physiology*, https://doi.org/10.1007/978-3-031-56296-9_17

The two ejaculatory ducts enter the prostate and join the **urethra** (distal to the internal urethral sphincter), which carries semen (or urine) out through the meatus of the penis.

3 ♂ Reproductive Physiology

Testes are kept at ~1.1 °C less than the body temperature (the cremaster muscle raises/lowers). The scrotal cavity lines with the tunica vaginalis (a serous membrane).

Testes produce **androgens** (testosterone) and "physically mature" spermatozoa. They provide ~5% of ejaculate.

Seminal vesicles produce alkaline seminal fluid (fructose, prostaglandins(!), ascorbic acid, fibrinogen), make sperm motile (flagellum begins to beat) and provide ~60% of ejaculate volume.

The prostate produces alkaline prostatic fluid containing citric acid, clotting enzymes and fibrinolysin to liquefy coagulated semen (20–30% of ejaculate into the urethra).

Bulbourethral (Cowper's) glands add <5% of alkaline lubricating fluid into the urethra.

4 Spermatogenesis

A 64-day process by which 46-chromosome germ cells, "spermatogonia", are transformed into 23-chromosome mature sperm "spermatozoa"

In seminiferous tubules:

One *stem cell* undergoes **mitosis** (division) to produce two cells, and one of these produces two more.

These two produce four *primary spermatocytes (46 chromos).*

Each primary spermatocyte undergoes **meiosis** → four *spermatids with 23 chromos (remain connected to each other and are surrounded by a nurse cell cytoplasm).*

Spermatids **mature** into *spermatozoa* (= sperm) by extruding cytosol and some organelles in a process called spermiogenesis.

Spermatozoa are **capacitated** when mixed with seminal vesicle fluid and fluid from "peg" epithelial cells of the fallopian tube in the ♀ reproductive tract.

5 Semen

Ejaculate (2–5 ml) contains 30–150 × 10^6 spermatozoa/ml, enzymes, electrolytes, protein, fructose, lipids, vitamin C, carnitene, prostaglandins, water and mucus.

♂ Hormones
The gonadotropin-releasing hormone (GnRH) (produced in the hypothalamus) stimulates the anterior pituitary to release a follicle-stimulating hormone (FSH) and luteinising hormone (LH) (= gonadotropins).

LH targets interstitial (Leydig) cells (of the testes) to produce testosterone.
FSH targets nurse (sustentacular/Sertoli) cells (in the testes), which (in the presence of testosterone) promote spermiogenesis.

6 Testosterone

(Formed from cholesterol)

- Stimulates spermiogenesis (along with FSH)
- Affects the central nervous system (CNS) function (libido)
- Stimulates the metabolic rate
- Stimulates muscle and bone growth
- Establishes and maintains secondary sex characteristics (hair distribution, muscle mass, body size, fat deposits, thicker vocal cords, Adam's apple, thicker skin, body shape, larger hands and feet, behavioural effects)
- Maintains glands and organs of the reproductive tract

7 "Men's (XY) Health"

- Cryptorchidism
- Inguinal hernias, testicular cancer
- Prostatitis, prostate cancer
- Orchiectomy (surgical removal of the testicles)
- Infertility ($<20 \times 10^6$ sperm/ml), impotence
- X-linked genetic disease
- Vasectomy
- Klinefelter's syndrome = polysomy X (XXY – i.e. have two X chromosomes like female but "are male"), small penis, tall stature (treated with testosterone); incidence ~ 1/600–750

8 Female Anatomy

- It consists of two ovaries, a pelvic cavity, two fallopian tubes, a uterus, a cervical os, a vagina, two greater vestibular (Bartholin's) glands and a vulva.

- Bartholin's glands secrete lubricating mucus.

Note: The clitoris is the most anterior/ventral, the urethral opening is anterior to the vaginal opening and the anus is posterior/dorsal.

9 Female Reproductive Tract

The **fallopian tubes** move the ovum towards the uterus via the cilia and peristalsis over 3–4 days. Fertilisation occurs here.

The **uterus** is subdivided into the fundus, body and cervix (neck). The wall of the uterus has three layers: the serous perimetrium, the muscular myometrium (where the muscle cells expand and elongate as the uterus expands to accommodate the foetus) and the glandular endometrium (which receives blastocyst/sloughs off).

The **cervical canal** dilates to allow the passage of the baby through the cervical os and into the vagina).

10 ♀ Reproductive Physiology

Ovaries store primordial follicles (= ovum surrounded by a single layer of follicle cells) paused in "prophase I" of meiosis. These are present at birth.

The FSH targets granulosa cells, causing the growth of 6–12 primary follicles each month.

1. An ovum increases in size.
2. **Follicle cells** proliferate into several layers and are called **granulosa cells**.
3. Theca cells surround the follicle.

Thecal cells produce androgens, which diffuse to granulosa cells, which convert androgens (i.e. androstenedione) to oestrogens.

Granulosa cells secrete a fluid that accumulates in the **antrum** within the surrounding granulosa cells: secondary follicle → one follicle outgrows the others (they become atretic) and grows to 1–1.5 cm = mature (Graafian/vesicular/tertiary) follicle.

Ovulation: the tertiary follicle releases secondary oocytes – paused in "metaphase II" of meiosis (meiosis completes after fertilisation) – into the pelvic cavity.

Cilia on the fimbriae of the infundibulum beat and produce currents, which move the oocyte into the fallopian tube.

Fertilisation occurs in the fallopian tube and zygote implants in the uterine wall.

The uterus provides mechanical protection, nutritional support and waste removal for the embryo.

The vagina is a passage for the elimination of menstrual fluids, receives the penis and forms the birth canal.

Gestation = 38 weeks, ~3.2 kg.
Premature = 28–36 weeks, >1 kg.

11 ♀ Hormones of the Ovarian Cycle

The ovarian cycle is a monthly cycle of events associated with the maturation of the ovum – occurs in the ovary.

1. The hypothalamus releases GnRH (at 1 pulse/60–90 min).
2. GnRH causes FSH (and LH) release from the anterior pituitary.

(a) **Follicular (pre-ovulatory) phase (days 1–14, may vary)**

3. The FSH acts on the ovary to stimulate follicle development.
4. The developing follicle produces oestrogen (which inhibits LH release) and inhibin (which causes FSH levels to decline, i.e. −ve feedback).
5. So the blood oestrogen level rises, which causes the pulse frequency of GnRH release to increase to 1 pulse/30–60 min.
6. This increased GnRH and stimulates LH release from the anterior pituitary.
   This "surge" in LH at day 13 → the completion of meiosis I by the oocyte and ovulation and development of the **corpus luteum** (CL) (= granulosa cells of the tertiary follicle).

(b) **Luteal (post-ovulatory) phase (days 14–28)**

7. The corpus luteum releases progesterone – which prepares the uterus for pregnancy (and some oestrogen).
8. As the progesterone level rises and oestrogen falls, GnRH pulse frequency declines to 1–4 pulses/day.
9. This stimulates LH secretion (which maintains CL) more than it does FSH.
10. If pregnancy does not occur, CL deteriorates (after 10 days) and oestrogen and progesterone levels drop.
11. This drop allows the frequency of pulses of GnRH release to increase, which stimulates FSH release → next cycle.

12 Estradiol

Thecal cells produce androgens (formed from cholesterol via testosterone), which diffuse to granulosa cells, which convert androgen to oestrogen.

Oestrogen:

- Stimulates bone and muscle growth
- Maintains secondary sex characteristics (hair distribution, breasts, location of adipose tissue, voice, pelvis)
- Affects the CNS (sexual behaviour)
- Maintains accessory reproductive glands/organs
- Initiates the growth and repair of the endometrium
- Promotes the development of breasts

13 The Menstrual (Uterine) Cycle

- <u>Occurs in the uterus</u>
- ~28-day cycle
- The cyclical preparation of the uterus to receive a fertilised ovum and the subsequent destruction of tissue if no embryo implants

  (a) **Menses** (menstruation) (**days 1–7**) – endometrial sloughing; 35–50 ml of blood is lost.
  (b) **Proliferation (days 5–14)** – the uterine epithelium regrows under the stimulation of oestrogen (secreted by developing ovarian follicles).
  (c) **Secretory phase (days 15–28**) begins at ovulation and continues while the corpus luteum is intact (endometrial glands secrete glycogen – a nutrition source for embryo).

(The day of ovulation is determined by counting back 14 days from the first day of menses to the start of the luteal phase of the ovarian cycle.).

(Menarche, menstrual cycles, menopause)

14 "Women's Health" XX

- Ovarian cancer, cervical cancer
- Ectopic pregnancy
- Pelvic inflammatory disease, endometriosis
- Breast cancer, hysterectomy
- Menarche, menopause, amenorrhea
- Preeclampsia (= pregnancy-induced hypertension, albuminuria, oedema at 32 weeks+)
- Tubal ligation
- Salpingectomy (salpinx = tube)
- Turner's syndrome = monosomy (single) X (i.e. have one X chromosome like males but "are female"), short stature, no ovaries (no penis), no secondary sex characteristics; incidence: ~1/2500–10,000

15 Fertilisation

Sperm are deposited near the cervical os and so must pass through the cervical mucus. A less viscous mucus is produced when the oestrogen level is high (at ovulation), which allows sperm's easy passage. Thick mucus impedes sperm entry to the uterus.

Sperm reach the fallopian tube ~30 min after ejaculation and live for ~50 h.

Mature eggs release the sperm chemoattractant "allurin".

The acrosomal enzymes of many sperm are required to tunnel through the ovum's zona pellucida. The fusing of one sperm to an ovum's plasma membrane prevents the entry of any other sperm.

16 Genetics

Cells of ♀ have 46 chromosomes, including XX (22 + X from the mother and 22 + X from the father).

Cells of ♂ have 46 chromosomes (22 + X from the mother and 22 + Y from the father).

$$23 = n, \qquad 46 = 2n$$

In turn, each chromosome may be composed of one chromatid (after mitosis) or (after duplication) two identical chromatids.

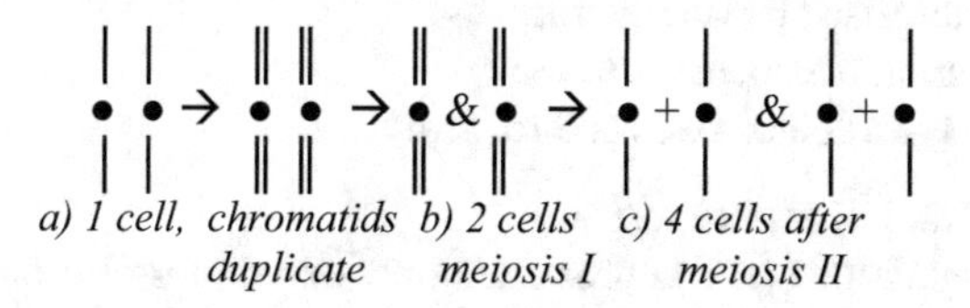

(a) A chromosome pair has one chromatid per chromosome. Then chromatids duplicate, and the chromosome pair (with two chromatids per chromosome) is called a tetrad (at metaphase I of a single cell).
(b) Each chromosome of the pair moves to a separate cell – we now have two cells.
(c) At anaphase II, the double chromatids of each chromosome separate into single chromatids, which become four separate cells (gametes).

The purpose of sexual reproduction is to produce genetic variability in individuals.

Spermatogenesis starts with the *mitosis* of a stem cell (which is $2n$, and each chromosome consists of two chromatids).

The chromatids of each chromosome detach and move apart to form two new cells (primary spermatocytes). New cells are $2n$, but each chromosome consists of one chromatid.

Each chromatid then duplicates itself so that each chromosome of the new spermatocyte again consists of two chromatids.

A primary spermatocyte undergoes *meiosis* to produce two new cells (secondary spermatocytes), each with n chromosomes of two chromatids (*the exchange of genetic material (crossing over) during the pro-phase of meiosis I introduces variability, and independent assortment during the meta-phase of meiosis I also introduces variability)*.

In each secondary spermatocyte, the chromatids separate and move apart again (to form a total of four cells – called spermatids – each with n chromosomes of one chromatid). (Two of these spermatids will contain an "X", while the other two will contain a "Y".)

Thus, each spermatid (and sperm) has 23 chromosomes (not 46).

Spermiogenesis – undifferentiated spermatids differentiate into mature spermatozoa (major changes in internal and external structures).

Oogenesis – the stem cell divides by mitosis into primary oocytes ($2n$) before birth. After puberty, the primary oocyte completes meiosis to produce (up to) three polar bodies and one ovum with 23 chromosomes.

Genetic variability (in the sperm and ovum) is introduced at synapsis (crossing over between adjacent chromatids) and at metaphase I (maternal and paternal chromosomes separate randomly).

Fertilisation – one sperm (with 23 chromosomes) enters the ovum (which contains 23 chromosomes) to form a zygote which has 46 chromosomes.

Zygote = ovum fertilised by one sperm
Pre-embryo (morula, blastocyst, <1 week)
Embryo (weeks 1–8), foetus (weeks 9 to term)

Oestrogens – estradiol, estrone, estriol
Androgens – testosterone (and androstenedione, dihydrotestosterone)
Gametes – spermatozoa, ovum
Gonads – testes, ovaries
Primary sex characteristics – gonads, external genitalia and reproductive tract
Secondary sex characteristics – structures that appear after puberty in response to sex hormones and are not involved in the production of gametes

17 Additional Information

Seminal fluid contains enzymes.
Protease dissolves vaginal mucus.
Seminal plasmin kills bacteria.
Prostatic enzyme converts fibrinogen to fibrin to coagulate semen.

Fibrinolysin liquefies clotted semen after 20–30 min.

The beginning (Harvard):

https://www.youtube.com/watch?v=ieYdgon-wT4

Androgens, especially dihydrotestosterone (DHT), mediate hair growth throughout the body and have opposite effects depending on the area of the body, increasing hair growth on the face, chest, pubic, axillae and extremities and inhibiting hair growth on the scalp in men who are genetically predisposed to male pattern hair loss or male androgenetic alopecia.

Prostate cancer is the most commonly diagnosed cancer in Australian men and is the second-leading cause of cancer mortality in Western men (lung cancer is first, and colorectal is third).

Twenty-nine percent (29%) of all diagnosed ca are prostate; 1/6 of men will be affected (mean age = 71 years).

(Treated with bilateral orchidectomy or luteinising hormone-releasing hormone-receptor agonists or prostatectomy, with seminal vesicles or radiotherapy)

In females, breast cancer is the first most frequent malignancy; colorectal is second.

Endometriosis is defined as the presence of endometrial-like tissue outside the uterus (often on the peritoneum). It affects 6–10% of women; symptoms = pelvic pain.

18 Reproductive System Revision: Homework Exercise 17

1. List the structures of the male reproductive tract in the order that a spermatozoon would pass through them.
2. What is the function of testosterone in males?
3. Where do spermatozoa physically mature, and where do they become "capacitated"?
4. What happens in the ejaculatory duct?
5. What is the composition of semen?
6. Write a paragraph that summarises the events of spermatogenesis. (Do not merely copy out the textbook!)
7. What is the function of estradiol?
8. Summarise the hormonal control of the ovarian cycle up to the luteal phase.
9. What are the male secondary sex characteristics?
10. What are the female primary sex characteristics?
11. Define menstrual cycle, menarche and menopause.
12. Outline the major changes in FSH, LH, estradiol and progesterone before and after ovulation.
13. What is the function of GnRH and the result of its release in males and females?

Suggested Answers to the Lecture Revision Homework Exercises

Exercise 1: Cells and Tissues

1. *Name four of the organelles in a cell and describe their function.*

   Mitochondria: produce ATP (95% of ATP required by the cell).
   Lysosomes: membrane-enclosed vesicles that contain enzymes capable of breaking down a variety of molecules; removal of damaged intracellular organelles or pathogens.
   Nucleus: contains DNA and associated protein (chromatin); control of protein synthesis and metabolism
   Endoplasmic reticulum: stores newly synthesised molecules; synthesise fatty acids, phospholipids and cholesterol. Rough ER: modifies and packages newly synthesised protein. Smooth ER: synthesise lipids and carbohydrates.
   Golgi complex: processes and delivers lipids and proteins to the plasma membrane for secretion.

2. *Describe the structure of the plasma membrane (cell membrane).*

   Double layer of phospholipid molecules – the lipid ends of the two layers of molecules face each other, while the phosphate heads face either the inside of the cell or the extracellular environment. Cholesterol molecules are scattered throughout the phospholipid molecules, and there are protein molecules embedded in the membrane as well – some of these form channels that allow the passage of some substances through the membrane. Others may act as receptor proteins that are sensitive to the presence of specific extracellular molecules, e.g. calcium, hormones, etc.

M. Caon, *Lecture Notes, Worksheets, and Exercises for Basic Anatomy and Physiology*, https://doi.org/10.1007/978-3-031-56296-9

3. *Define the processes of "diffusion" and "osmosis".*

   Diffusion is the random movement of molecules (particles) within a gas or liquid. The effect of diffusion is to eventually distribute particles from regions where they are in high concentration to regions where they are in low concentration.
   Osmosis is the diffusion of water molecules through a semi-permeable membrane from one solution to another solution that contains a higher solute concentration.

4. *What roles do proteins play in a cell's plasma membrane?*

   Some proteins are enzymes (they make chemical reactions happen faster), some are receptor proteins (for signalling chemicals, e.g. hormones), some are channels that transport molecules and ions (selectively allow entry to some solutes, e.g. ions), some are anchoring proteins that act as attachments (to cells' cytoskeleton and to extracellular structures) or join cells to adjacent cells (e.g. desmosomes), some are recognition protein used to identify the cell (glycoproteins = identification tags) and some are carrier proteins that bind solutes and transport them across the plasma membrane.

5. *What is active transport?*

   It is the movement of particles against their concentration gradient across the plasma membrane using ATP.

6. *Name and briefly describe the four types of tissue.*

   Epithelial: covers body surfaces (skin), lines the interior or exterior of hollow organs and ducts and forms glands.
   Connective: connects the epithelium to the rest of the body and protects (cushions and immunity) and supports (bone and cartilage) the body and organs; consists of specialised cells and an intercellular matrix. Other types are blood, bone and fat.
   Muscle: produces movement by contracting. Three types: skeletal, smooth and cardiac muscle.
   Nervous: the brain and spinal cord and peripheral nerves. Gray matter: nerve cell bodies; white matter: axons.

7. *What are the functions of epithelial tissue?*

   (a) Physical protection: it lines the cavities of hollow organs, ducts and body cavities and covers the body – places that are likely to dehydrate or lose tissue from abrasive action.
   (b) Controlling permeability: every substance that enters or leaves the body must cross an epithelium.
   (c) Sensation: most epithelia have a large nerve supply and hence are extremely sensitive to stimulation.
   (d) Providing specialised secretions: it makes up the bulk of glands and produce secretions.

8. *What is the difference between "loose" connective tissue and "dense" connective tissue?*

   Loose connective tissue (adipose tissue, synovial membrane): much of the space is occupied by "ground substance".
   Dense connective tissue (tendons, ligaments, heart valves, dermis) – fibres occupy more space than cells or ground substance.

9. *From what type of tissue are the following structures made: bone, lymph, tendon, cartilage, adipose tissue, glands, epidermis?*

   Bone: connective tissue.
   Lymph: a liquid; lymph tissue is connective tissue.
   Tendon, cartilage: both are dense connective tissue.
   Adipose tissue: loose connective tissue.
   Glands, epidermis: both are epithelial tissue.

10. *What structure separates the thoracic and abdominal cavities, and what is it made of?*

    The diaphragm. It is made of skeletal muscle (it can be under our conscious control).

11. *What is the collective name for the contents of the ventral cavity*

    The viscera

12. *What are the main functions of these membranes and the potential space they form?*

    To secrete serous fluid, which allows the parietal membrane to slide over the visceral membrane without friction.

13. *What is the clinical condition that develops when air is able to enter the potential space of the pleural membrane?*

    Pneumothorax.

14. *What is the clinical condition called when the membrane of the abdominal cavity is inflamed?*

    Peritonitis.

15. *How does an organ differ from a tissue?*

    Tissues *are collections of specialised cells and cell products that perform a relatively limited number of functions (or a specific function) [or one or more specific functions].*
    Organs *are combinations of tissues that perform complex functions (or two or more tissues working in combination to perform several functions).*

16. *Using the gastrointestinal tract as an example, list the cavity/cavities in which organs of this system are found.*

    (Buccal cavity), thoracic cavity (oesophagus), abdomino-pelvic cavity (rest of organs) – may separate organs into abdominal cavity and pelvic cavity.

17. *Do all organs of the body lie within a body cavity? If not, give examples.*

    No. Muscles and bones lie outside cavities; kidneys (and part of the pancreas and duodenum) are "retroperitoneal", so they lie "outside" the peritoneum. The abdominal cavity encloses the organs surrounded by the peritoneum and retroperitoneal organs.

18. *Using directional terms, describe the appearance of the body when it is standing in the "anatomical position".*

    Body vertical, feet inferior to the waist and knees, feet not everted, legs neither flexed nor extended, arms lateral and parallel to the trunk, hands inferior to the elbows, face directed anteriorly, neck neither flexed nor extended.

19. *Describe the position of each of the following using anatomical, directional terms:* ***ear*** *(compared to the nose and to the chin),* ***elbow*** *(compared to the wrist and shoulder) and* ***vertebrae*** *(compared to the sternum and kidneys).*

    The ear is lateral to the nose and superior to the mandible.
    The elbow is proximal to the wrist but distal to the shoulder.
    Vertebrae are posterior (or dorsal) to the sternum and medial to kidneys.

Exercise 2: Basic Chemistry Revision

1. *Write out the definition of each of the following 15 things:*

   Element: the 90 naturally occurring simplest substances (listed in the "periodic table" – have *chemical symbols*).
   Atom: the smallest particle of an element (contains protons, neutrons and electrons).
   Proton: +vely charged subatomic particle in the nucleus of an atom (atoms of different elements have different numbers of protons).
   Neutron: subatomic particle in the nucleus of an atom.
   Electron: –vely charged subatomic particle outside of the nucleus; tiny yet occupies the bulk of space in an atom.
   Chemical bond: the outer electron(s) of an atom participate with those of another atom in joining two or more atoms together to form a new substance (a "compound").
   Metal elements (LHS of the periodic table): always lose (donate) electrons in chemical reactions.

Non-metal elements (RHS of the periodic table): always gain electrons in chemical reactions.
Compound: a substance formed when atoms from two or more elements are chemically combined in fixed proportions (have a formula, e.g. H_2O, $C_6H_{12}O_6$).
Molecular compounds: atoms forming the molecule are from non-metal elements, covalently bonded.
Covalent bond: the bond between two *non-metal* atoms in a molecule (the atoms share electrons, so BOTH *gain* electrons).
Molecule: the smallest particle of a molecular compound; consists of two or more atoms joined by covalent bonds.
Ionic (non-molecular) compounds: formed when metal atoms (ionically) bond to six to eight surrounding non-metal atoms (and vice versa) – continuous crystal lattice structure.
Ionic bond: the attraction between a metal atom and **all** the surrounding non-metal atoms in the lattice (the non-metal atoms gain electron(s), while the metal atoms lose electron(s)).
Ion: an atom that has gained (if a non-metal) or lost (if a metal) one (or more) electron(s) (when an ionic substance dissolves in water, ions separate and move about freely as **electrolytes**). Molecules can be ions too!

2. *Identify the name of the elements and the number of atoms in the following:*

(a) $C_6H_{12}O_6$	Carbon (six atoms), hydrogen (12 atoms), oxygen (six atoms)
(b) CH_3COOH	Carbon (two atoms), hydrogen (four atoms), oxygen (two atoms)
(c) NH_4^+ OH^-	Nitrogen (one atom), hydrogen (five atoms), oxygen (one atom)

3. *Which of the following compounds are covalent, and which are ionic?*

(a) $C_6H_{12}O_6$	covalent
(b) CH_3COOH	covalent
(c) Na^+ Cl^-	ionic
(d) K^+ Cl^-	ionic
(e) Ba^{++} SO_4^{--}	ionic

Chemistry Calculations

1. By following the steps below, calculate the mass of substance per 100 ml for an IV solution that contains 0.3% Na^+Cl^- and 3.3% glucose.
   (a) Consider sodium chloride first. What mass of sodium chloride is in 100 ml of 0.3% Na^+Cl^-? **100 ml × 0.3% (g/ml) = 0.3 g.**
   (b) Repeat the procedure for glucose. What mass of glucose is in 100 ml of 3.3% glucose? **100 ml × 3.3% = 3.3 g.**
   (c) Now add the two masses together to get a total mass per 100 ml of 0.3% Na^+Cl^- and 3.3% glucose. 0.3 g + 3.3 g = 3.6 g.

2. *Calculate the mass of substance per 100 ml for a 0.224% K^+ Cl^- solution.*

   The mass of KCl: 100 ml × 0.224% = 0.224 g.

3. *Calculate the number of millimoles per litre in a 5% glucose solution.*

   (a) First calculate the mass of glucose in 1 l: mass (g) in 100 ml = **5** g.
   (b) Multiply this answer by 10 to get grams per litre: **5** ×10 = **50** g/l.
   (c) Calculate the number of moles in this many grams of glucose in two steps:

      (i) Number of grams in one mole of glucose $C_6H_{12}O_6$ (RAMs are C = 12, H = 1, O = 16). Multiply the RAM values by the number of atoms of each element appearing in the formula:

$$(12\times\underline{\mathbf{6}})+(1\times\underline{\mathbf{12}})+(16\times\underline{\mathbf{6}})=\underline{\mathbf{72}}+\underline{\mathbf{12}}+\underline{\mathbf{96}}=\mathbf{180}$$

   The answer is the "relative formula mass".

      (ii) Divide the number of grams per litre [from b) above] by the relative formula mass [from c) part (i)].

      The answer is the number of moles per litre (it should be a decimal less than 1.0). **50 ÷ 180 = 0.278 mol/l.**

   (d) Multiply the answer in (c) part (ii) by 1000 (shift the decimal three places right) to get the answer.

$$0.278\,\text{mol}/1\times1000=278\,\text{mmol}/1.$$

4. *Calculate the number of millimoles per litre in a 1 l IV bag of normal saline (0.9% Na^+ Cl^-).*

   1000 ml × 0.9 g/ml = 9 g. Relative formula mass of NaCl = **23 + 35.5 = 58.5 g/mol.**
   Number of moles/l = **9 g ÷ 58.5 g/mol = 0.154 mol = 154 mmol.**

5. *Give the definition of an osmole.*
The amount of substance that must be dissolved to produce 1 mole (which is 6.02×10^{23} molecules) of separate solute particles

   (a) What are the particles in the glucose solution? **Molecules of glucose.**
   (b) What are the particles in the saline solution? **Sodium ions and chloride ions.**

6. *State the number of moles per litre calculated in Q3 to state the number of osmoles per litre in 1 l IV bag of 5% glucose (a covalent molecular substance ← hint!)*

$$278\,\text{mmol}/1=0.278\,\text{mol}/1=0.278\,\text{osmol}/1.$$

7. *Followed by Q4, calculate the number of osmol/l in a 1 l bag of 0.9% Na^+ Cl^- (normal saline – an ionic substance)*

$$154\,\text{mmol}/\text{l} = 0.154\,\text{mol}/\text{l}.$$

$$0.154\,\text{mol}/\text{l} \times 2\left(\text{dissolved particles: Na} + \text{and Cl}^-\right) = 0.308\,\text{osmol}/\text{l}.$$

8. *Characterise the following solutions as hypotonic, isotonic or hypertonic to blood plasma:*

   (a) 0.9% sodium chloride – **isotonic**
   (b) 0.3% sodium chloride – **hypotonic**
   (c) 5% glucose – **isotonic**
   (d) 4% glucose – **hypotonic**
   (e) A solution containing 3.3% glucose and 0.3% sodium chloride – **isotonic**
   (f) A solution containing 4% glucose and 0.18% sodium chloride – **isotonic**
   (g) 9% sodium chloride – **hypertonic**

9. *(a) What is the pH of a solution that contains H+ at a concentration of 10^{-5} mol/l?*

$$\text{pH} = -\log\left(c\left(\text{H}+\right)\right) = -\log\left(10^{-5}\right) = 5.$$

(b) Is the solution in (a) acidic or basic? Acidic.
(c) What is the pH of a solution that contains H^+ at a concentration of 3.2×10^{-5} mol/l?

$$\text{pH} = -\log\left(c\left(\text{H}^+\right)\right) = -\log\left(3.2\times10^{-5}\right) = 4.49.$$

Exercise 3: Integument

1. *List the components of the integument.*
   Two major parts – the cutaneous membrane and the accessory structures:

   (a) Cutaneous membrane: epidermis (stratum basale, spinosum, granulosum, lucidum – thick skin only; corneum – on the outside) and dermis (reticular – on the inside – and papillary dermal layers).
   (b) Accessory structure: hair, exocrine glands and nails; they are located in the dermis and protrude through the epidermis to the skin surface.

2. *What are the principal functions of the skin?*

   (a) Body temperature regulation (either insulation or evaporative cooling).
   (b) Protection of underlying tissues and organs against impact, abrasion (production of keratin), dehydration, UV (production of melanin) and chemical attack.

(c) General sense, detection of touch, pain, perception of stimuli etc., which provide information on the external environment.
(d) Excretion of salt, water and organic wastes by integumentary glands.
(e) Synthesis of vitamin D, which is converted to calcitriol (a hormone that is important to calcium metabolism).
(f) Immune response to pathogens and cancers in the skin.

3. *Describe the five layers of the epidermis.* (**Give a description similar to that supplied in the lecture handout.)

   (a) Stratum basale: the deepest basal layer; attachment to the underlying dermis by protruding epidermal ridges; single-cell thick; contains basal cells, which push up the older ones above, tactile cells that are sensitive to touch and melanocytes.
   (b) Stratum spinosum: eight to ten layers of keratinocytes bound together by desmosomes.
   (c) Stratum granulosum: three to five layers of keratinocytes derived from the stratum spinosum and accumulate keratin and keratohyalin; keratohyalin promotes waterproofing.
   (d) Stratum lucidum: appears as a glassy layer in thick skin only (palm and sole), flattened dense cells filled with keratin and without organelles.
   (e) Stratum corneum: the most superficial layer; 15–30 layers; 75% of the thickness of the epidermis; dead, flat, durable, expendable; filled with keratin; water resistant but not waterproof; permits slow water loss by insensible perspiration.

4. *List the types of cells in the epidermis, how do they differ?*

   (a) Basal cells (germinative cells): stem cells that divide to replace the more superficial keratinocytes.
   (b) Keratinocytes: produce **keratin** (a fibrous protein).
   (c) Melanocytes: produce melanin, a pigment that protects against UV.
   (d) Dendrocytes (Langerhans' cells and Granstein cells): interact with lymphocytes (T cells) to assist in the immune response; are macrophages.
   (e) Merkel cells (tactile cells): associated with a sensory nerve ending.

5. *Describe the structure of the two strata in the dermis.* (**Give a description similar to that supplied in the lecture handout.)

   (a) Papillary layer: has folds and ridges called papillae that push up into the epidermal layer; contains capillaries, lymphatic vessels and sensory neurons that supply the skin.
   (b) Reticular layer: 80% thickness of the dermis, contains Pacinian corpuscles sensitive to deep pressure, contains both collagen fibres and elastic fibres, gives skin strength and resilience.

6. *What are the two types of glands in the skin, and what is their purpose?*

(a) Sebaceous (oil) glands: they secrete sebum into hair follicles and keep the skin and hair from drying out. Sebum is bactericidal.
(b) Sudoriferous (sweat) glands: apocrine sweat glands; sweat glands in the axillary, nipple and anogenital areas. The ducts produce sweat and fatty substances and proteins into hair follicles. Merocrine sweat glands: coiled tubes in the dermis discharge their secretions directly onto the surface of the skin. They are far more widely distributed than apocrine sweat glands.

Exercise 4: Thermoregulation

1. *Define what heat is.*

   When two objects are close to each other, the one at the higher temperature will pass energy to the one at the lower temperature. The energy that is transferred due to the temperature difference is known as heat.

2. *Define what temperature is.*

   The temperature of an object is the measure of the average kinetic energy of the particles of the object. That is, both the speed of the particles and their mass are involved. All particles are in constant motion (or constantly vibrating); the greater is the average KE, the greater is the temperature of the object.

3. *What are the means by which the body can lose heat?*

   Heat energy is lost from the body by (a) the evaporation of sweat (and water from the lungs), (b) radiation of infrared waves, and (c) convection and conduction to air or other objects (also by eliminating and urinating).

4. *Outline the skin's role in temperature regulation.*

   Heat is radiated through the skin surface – the greater the surface area, the greater the amount of heat lost.
   Heat is lost by convection from the skin – the greater the exposed area of the skin, the greater the amount of heat lost.
   Heat is lost due to the evaporation of sweat from the skin surface when sweat is released under the influence of the sympathetic division of the autonomic NS.

5. *Explain how sweating allows the body to lose heat.*

   When sweat evaporates, water is changing from liquid form to gaseous form. For this to happen, water molecules must separate themselves from the attractive forces of their neighbours. They cannot do this unless they are travelling quite fast, that is, have more than the average kinetic energy. Some of the water molecules are travelling fast enough and do evaporate. This means the more energetic water molecules escape from the skin (taking their kinetic energy with them), thus leaving the skin cooler by virtue of the fact that some energy (heat) has left the body surface. The fastest molecules are at a higher temperature, so the molecules left behind have a lower average kinetic energy – i.e. are cooler than the evaporated ones. This cooler sweat can then gain heat from the body as blood transfers it to the surface.

6. *Define what is meant by homeostasis.*

   The physiology of the body is in a dynamic state of equilibrium; internal conditions change and vary (oscillate) within relatively narrow limits. Homeostasis is the body's automatic tendency to maintain a relatively constant (i.e. within a narrow range) internal environment. This includes maintaining a relatively constant temperature, blood pressure, ion concentration, pH, hydration, dissolved gas concentration and food molecules and the prevention of the accumulation of wastes (among others).

7. *Explain how negative feedback maintains homeostasis.*

   Homeostasis returns the body to a healthy state after stressful stimuli by biofeedback mechanisms. "Negative" feedback means that the body's response is to make a change that opposes the stress (i.e. moves the body's physiology in the opposite direction to the change detected by a receptor).
   A receptor receives a stimulus about a variable and sends a message via an afferent pathway to an integrating centre. The integrating centre determines the normal level of the variable – the "set point" – and if the stimulus indicates a move away from the set point, a message is sent via an efferent pathway to the effector organ. The effector organ produces a response that moves the variable value back towards the wet point.

Exercise 5: Musculo-Skeletal System

1. *List the functions of the skeletal system.*

   The skeletal system has five primary functions:

   (a) Support: it provides structural support for the entire body and a framework for the attachment of soft tissue and organs.

   (b) Minerals and lipid storage: calcium is the most abundant mineral in the human body. The calcium salts of bones maintain the concentrations of calcium and phosphate ions in body fluids; meanwhile, bones store energy as lipids in areas with yellow bone marrow.

   (c) Blood cell production: RBC, WBC and platelets are produced in the red bone marrow.

   (d) Protection: it surrounds many soft tissues and organs. For example, the rib protects the heart and lungs; the skull encloses the brain.

(e) Leverage and movement: it functions as levers that can change the magnitude and direction of force generated by skeletal muscles.

What are the two major divisions of the skeletal system?

(a) The axial skeleton forms the longitudinal axis of the body (the skull, vertebral column, thoracic cage).

(b) The appendicular skeleton includes the bones of the limbs and the supporting bone girdles that connect them to the trunk.

2. *How does the compact bone differ from the cancellous bone?*

   Compact bone is constructed of osteons (haversian systems); there are concentric layers (lamellae) between which are scattered osteocytes (one osteocyte per lacunae). Lamellae surround the central canal, which contains blood vessels. The osteocytes in lacunae are linked to each other and the central canal by canaliculi. Compact bone is denser than cancellous bone (spongy bone), which contains trabeculae but has no osteons. Trabeculae are a meshwork of bone struts with spaces in between.

3. *Describe the histology of bone (and name the structures).*

   Compact bone is constructed of osteons. An osteon consists of a central canal or haversian canal (which contains blood vessels) surrounded by concentric lamellae of bone. Between lamellae are spaces (called lacunae) that contain bone cells (osteocytes). Between lacunae and the haversian canal are narrow corridors (called canaliculi) through which osteocytes extend "processes" in order to obtain nourishment.

4. *Characterise the following types of joint (symphyses, syndesmoses, synostosis, synchondroses, gomphoses, sutures, hinge, and ball and socket) according to:*

   (a) *Structure: as fibrous, cartilaginous or synovial*

   Fibrous joints = syndesmoses, gomphoses, sutures
   Cartilaginous joints = synchondroses, symphyses
   Synovial joints = hinge, ball and socket
   Bony = synostosis (fusion of the epiphysis and the diaphysis at the cessation of ling bone growth)

   (b) *Function as synarthroses = sutures, synchondroses, gomphoses, synostosis*

   Amphiarthroses = syndesmoses, symphyses
   Diarthroses = hinge, ball and socket

5. *Give an example of each of the following types of bone: long, short, flat, irregular and sesamoid.*

   Long: femur (and many others, e.g. arm, forearm, thigh, leg, palms, soles, fingers and toes etc.)
   Short: any carpal or tarsal

Flat: all cranial bones, sternum, ribs
Irregular: vertebrae, os coxae (hip bone)
Sesamoid: patella

6. *Perform the following movement with your left upper extremity: with the thumb and third and fourth fingers fully flexed and the first and second fingers fully extended, circumduct the arm in a flexed position with the lower arm supinated.* (Find out how to do it!)
7. *Compare AND contrast the anatomy of the knee joint and the elbow joint. In what way(s) does(do) their structure reflect their function?*

Both are synovial joints and hinge joints and involve more than two bones. Both have collateral ligaments for stability. The knee involves a fourth bone – the patella; it is weight bearing, so it has a greater area; has menisci (fibrocartilaginous pads) to cushion and extend the articulating surface and has "cruciate" ligaments within the joint capsule.

8. *Find an example of a superficial muscle that is descriptively named according to each of the eight types of descriptors: direction of muscle fibres, location in the body, relative size, number of origins, shape, origin and insertion, action and whimsy.*

- Direction of muscle fibres: rectus femoris
- Location in the body: erector spinae
- Size: pectoralis major
- Number of origins: biceps femoris
- Shape: serratus anterior
- Origin and insertion: sternohyoid (O = manubrium, I = hyoid bone)
- Action: extensor hallucis longus (extends the big toe)
- Whimsy: soleus (resembles the flat fish known as the "sole"!)

9. *Choose six muscles (different from those in Q 8), one from each of the lower arm and leg, upper arm and leg, and front and back of the body, and list their origin O, insertion I and action A.*

For example (for the anatomical position):

Lower arm: flexor carpi radialis, O = medial epicondyle of the humerus, I = second + third metacarpal bones, A = flexion of the wrist (see Table 11.13 in Martini 3rd ed.).
Forearm: flexor carpi ulnaris, O = medial epicondyle of the humerus; I = pisiform, the hook of hamate and the base of the fifth metacarpal; A = flexes and adducts, hand at wrist.
Leg: tibialis anterior, O = upper 1/2 and lateral condyle of the tibia, I = medial cuneiform and first metatarsal bones of the foot, A = dorsiflexion and inversion of the foot.
Arm: biceps brachii, O = long head (passes over the top of the joint laterally to the scapula), I = via a strong tendon to the radial tuberosity and bicipital apo-

neurosis, A = strong flexor and supinator of the forearm, also weaker effects of abducting (long head) and adducting (short head) the shoulder.
Thigh: rectus femoris, O = straight head from the anterior inferior iliac spine, reflected head from groove just above acetabulum; I = inserts into the patellar tendon as one of the four quadriceps muscles; A = knee extension, hip flexion.
Anterior of the body: pectoralis major, O = clavicular head, sternal head; I = lateral lip of the intertubercular (bicipital) groove of the humerus, crest of the greater tubercle of the humerus, A = clavicular head (flexes the humerus), sternocostal head (extends the humerus).
Posterior of the body: trapezius, O = external occipital protuberance, nuchal ligament, medial superior nuchal line, spinous processes of vertebrae C7-T12; I = posterior border of the lateral third of the clavicle, acromion process and spine of the scapula; A = rotation, retraction, elevation and depression of the scapula.

10. *Describe the relationship between the three proteins of the thin myofilaments – actin, troponin and tropomyosin – and how they interact.*

    Tropomyosin twines around an actin molecule so that tropomyosin covers the sites on actin, to which myosin molecules can bind to produce muscular contraction.
    Troponin is attached to tropomyosin.
    Troponin changes shape when a Ca^{++} ion binds to it and as a consequence shifts tropomyosin away from the binding sites of actin – i.e. the sites at which myosin binds to actin are now exposed. Once the active sites are exposed, the energised myosin heads bind to them, forming a cross-bridge, and the contraction cycle begins.

11. *In which directions do the muscle fibres in the external obliques and rectus abdominus muscles lie?*

    Obliques run at an angle to vertical (not vertical and not horizontal).
    Rectus fibres are vertical (if the person is standing, hence are “erect”).

12. *Comment on the relative size of the gluteus maximus and the gluteus medius.*

    Maximus – as the word implies – is the bigger of the two.
    Medius is the smaller of the two (implies that there is a minimus!).

13. *Which muscles extend the spine, and which muscles extend the arm?*

    Erector spinae (spine), triceps brachii (arm).

14. *What actions do the following muscles perform?*

    (a) Pronator quadratus – pronates the forearm.
    (b) Flexor carpi ulnaris – flexes the wrist.
    (c) Extensor digitorum – extends the fingers.
    (d) Extensor carpi ulnaris – extends the wrist.

15. *Over which bones do the extensor carpi ulnaris and the extensor carpi radialis brevis lie?*

    The ECU is over the ulna; the ECR is over the radius.

16. *What part of the name biceps femoris indicates its location?*

    The femoris part = along the femur.

17. *How many origins do the triceps brachii have, and where are they?*

    Three – the tri- part of the name tells us this. They are infraglenoid tubercule of the scapula, posterior shaft of the humerus and posterior humeral shaft distal to the radial groove.

18. *What is the origin and insertion of the sternocleidomastoid muscle?*

    Origin = sternum (the manubrium part) and clavicle.
    Insertion = mastoid process of the temporal bone.

19. *What broad general shapes do the a) rhomboid major muscle and b) deltoid muscle resemble?*

    A rhombus (tilted square) and a capital Greek letter delta, respectively.

20. *What action is performed by the masseter?*

    It elevates the mandible (i.e. gritting teeth, biting, chewing food) – by using this muscle we "masticate" food.

21. *Under which gluteal muscle does the sciatic nerve lie?*

    Gluteus maximus. Hence, into which gluteal muscle do you insert the needle for IM injection?

22. *List the muscles that comprise the quadriceps group. Where are they?*

    (Quadriceps implies four muscles!) They are rectus femoris, vastus intermedius, vastus lateralis and vastus medialis. They can be found on the anterior surface of the thigh. Note that the three vastus muscles are on the medial surface or the lateral surface or between (i.e. intermediate) the two. Rectus femoris is over the femur and has muscle fibres that are "vertical" (if you are standing).

23. *List the muscles that comprise the hamstrings group.*

    Hamstrings lie on the dorsal part of the thigh. They are biceps femoris (lateral), semimembranosus (medial-ish) and semitendinosus (central).

24. *Where are the locations of the triceps brachii and the triceps surae?*

    The triceps brachii is on the posterior surface of the arm; the triceps surae is on the posterior surface of the leg (= gastrocnemius, which has two heads + soleus muscle).

25. *Give five reasons why the deltoid, vastus lateralis and gluteus medius muscles are chosen as the sites for intramuscular injection.*

    (a) They are easily accessible.
    (b) Large thick muscles can take larger quantities of drug than sub-cut or intradermally.
    (c) The drug will be absorbed gradually into the bloodstream (more gradually than for IV injection).
    (d) They are well vascularised with extensive fascia, so they have large surface areas for drug absorption.
    (e) They are situated away from major nerves and arteries – so less chance of accidental puncture.
    (f) Vastus lateralis is the site of choice for children (who have small deltoid and gluteal), and for the elderly who have atrophied muscles.

Exercise 6: Digestive System

1. *List the organs of the digestive system and the accessory organs.*

   Gastrointestinal tract (GIT) organs: mouth, pharynx, oesophagus, stomach, small intestine, large intestine and anus.
   Accessory organs: teeth, tongue, and various glandular organs, salivary glands, liver, gall bladder, pancreas

2. *Name the three regions of the small intestine and describe what digestive and absorptive events happen in each region.*

   Duodenum, jejunum and ileum.
   Most digestion and absorption happens in the duodenum and jejunum. The **duodenum** is the site of chemical digestion, a mixing bowl. It receives chyme from the stomach and digestive secretions (mucus) from the pancreas and liver. Iron and calcium are absorbed in the duodenum. The bicarbonate-rich mucus (from mucus-secreting duodenal glands in the sub-mucosa) neutralises stomach acid and raises pH to ~8. The duodenum receives bile from the gall bladder (it emulsifies fats), enzyme-containing pancreatic juice from the pancreas and "brush border" enzymes from the microvilli.
   In the **jejunum**, the bulk of chemical digestion and nutrient absorption occurs. It receives "intestinal juice" from intestinal crypts. Virtually all foodstuffs, 80% of electrolytes and most of the water are absorbed in the small intestine (before the ileum).
   The **ileum** is the final segment of the small intestine and is also the longest. The ileum's major absorptive role is to reclaim water and bile salts for recycling to the liver to be secreted in bile again. The ileum also absorbs vitamin B12 intrinsic factor.

3. *Describe the functions of the four tunics of the wall of the alimentary canal.*

   (a) **Mucosa**: secretes mucus, digestive enzymes and hormones; absorbs end products of digestion; protects against infectious disease (lymphoid nodules) and self-digestion.

(b) **Sub-mucosa**: dense connective tissue containing blood vessels, lymphatics (transport products of digestion) and nerve plexus (innervate gut). In some regions, it also contains exocrine glands, which secrete buffers and enzymes into the lumen of the digestive tract.
(c) **Muscularis externa**: smooth muscle cells dominate this region. Gut motility: peristalsis (propulsion) and segmentation (mixing). Inner circular muscle layer (→sphincters), outer longitudinal muscle layer. These layers play an essential role in mechanical processing and peristalsis (moving materials along the digestive tract).
(d) **Serosa (visceral peritoneum)**: areolar connective tissue covers the muscularis externa along most portions of the digestive tract inside the peritoneal cavity. The adventitia – fibrous connective tissue covering the muscularis externa of the oral cavity, pharynx and oesophagus rectum – attaches the gut to adjacent structures.

4. *Draw up a table with three columns and ten lines (make up your own headings) that lists (page 950) (a) each digestive enzyme, (b) which structure/cells secrete the enzyme and (c) what the enzyme does.*

(a) Each digestive enzyme	*(b) Which structure/cells secrete the enzyme*	*(c) What the enzyme does*
Salivary amylase	Salivary gland	Starch hydrolysis
Pancreatic amylase	Pancreas	Starch hydrolysis
Pancreatic lipase	Pancreas	Digest fats and oil
Pancreatic nuclease	Pancreas	Digest nucleic acids
Lactase, maltase, sucrase	Intestinal brush border	Lactose, maltose, sucrose hydrolysis
Pepsin	Gastric glands	Proteins hydrolysis
Dextrinase, glucoamylase	Intestinal brush border	Oligosaccharides hydrolysis
Chymotrypsin	Pancreas	Protein hydrolysis (proteolysis) to amino acid and small peptide
Carboxypeptidase, aminopeptidase, dipeptidase	Intestinal brush border	Protein hydrolysis
Nucleosidases, phosphatases	Intestinal brush border	Nucleic acid hydrolysis

5. *Describe all the processes that constitute mechanical digestion.*

Mastication: it involves using the teeth and tongue to break food into small pieces, crushing it and mixing it with saliva in the mouth to form a bolus for swallowing.
Swallowing (deglutition): the pharynx is a passageway between the oral cavity and the oesophagus. It has two layers of skeletal muscles to propel food into the oesophagus (deglutition). The oesophagus is a muscular tube that conducts solids and liquids from the pharynx to the stomach.
Stomach churning: in the stomach, peristaltic waves (3 per min) mix contents with gastric juice and squirt about 3 ml of liquid chyme through the pyloric sphincter into the duodenum.

Segmentation: this refers to an activity that mixes food with digestive enzymes and moves food backwards and forwards over the intestinal wall.

6. *Summarise the regulation of gastric secretion.*

   (a) **Cephalic (reflex) phase**: it involves thought, sight, smell and taste of food. This is directed by the CNS – a conditioned reflex that increases gastric secretion (before the food enters the stomach). This phase only lasts minutes.
   (b) **Gastric phase**: the stimuli that initiate the gastric phase are as follows: (1) stomach distension increases secretion; (2) the presence of undigested materials in the stomach, especially proteins, peptides (partially digested proteins) and caffeine, stimulates the release of gastrin (a hormone that stimulates the secretion of HCl); (3) a rise in pH level (indicating not enough acid in the stomach) stimulates the release of gastrin. This phase may continue for 3–4 h.
   (c) **Intestinal phase Excitatory**: food entering the duodenum stimulates the release of *intestinal gastrin*, which again stimulates gastric secretion (briefly).

7. *Summarise the events in the chemical digestion of carbohydrates.*

   Chemical digestion involves the *hydrolysis* of carbohydrates. The products are monosaccharides. This includes two steps: one step involves carbohydrases produced by the salivary glands and pancreas. The other step uses brush border enzymes. Carbohydrates are digested into disaccharides and trisaccharides by two enzymes: salivary amylase and pancreatic alpha-amylase. Brush border enzymes (dextrinase, glucoamylase, lactase, maltase, sucrase) of the intestinal microvilli covert disaccharides and trisaccharides into monosaccharides prior to absorption.

8. *Summarise the events in the chemical digestion of proteins.*

   The chemical digestion of protein begins in the stomach through the action of HCl, which disrupts the tertiary and secondary protein structure, exposing peptide bonds to enzymatic attack. Pepsin (proteolytic enzyme) in the stomach severs the peptide bonds within a polypeptide chain. When entering the duodenum, proteins are digested with the aid of trypsin and chymotrypsin (from the pancreas), carboxypeptidase, aminopeptidase and dipeptidase (from the brush border). Chemical digestion involves the *hydrolysis* of proteins and peptides. The products are amino acids. The pancreas is stimulated to release enzymes by parasympathetic nerves and by **cholecystokinin** (CCK), a local hormone released when protein and fats enter the duodenum. The release of HCO_3 is stimulated by **secretin** (a hormone released by the presence of acid in the duodenum).

9. *Summarise the events in the chemical digestion of lipids.*

This involves lingual lipase from glands of the tongue and pancreatic lipase from the pancreas. Lipids (triglycerides and their breakdown products) are insoluble in water. In the duodenum, where most lipid digestion happens (80%), the fat globules that they form are emulsified by the detergent action of bile salts. The emulsified fatty droplets are digested with the aid of pancreatic lipases. Chemical digestion involves the *hydrolysis* of lipids. The products are free fatty acids and monoglycerides.

10. *Describe the basic anatomy of the liver (the functional unit and the cells).*

The functional unit is the lobule (1 mm diameter) – a six-sided structure with a central vein and a portal triad at each "corner". The portal triad consists of the hepatic artery, hepatic portal vein and bile duct. Blood from the hepatic portal vein enters sinusoids (capillaries without a basal lamina) and mixes with arterial blood before draining into the central vein. Sinusoids contain Kupffer cells (macrophages). Plates/lines of hepatocytes radiate from the central vein. Bile drains from these via canaliculi, which connect with fine bile ductules, and they carry bile to the bile ducts in the nearest portal area.

11. *State the functions of the liver.*

(a) Carbohydrate metabolism: the liver stabilises blood glucose levels; it also synthesises glucose from other carbohydrates or from available amino acids (gluconeogenesis).
(b) Lipid metabolism: stores triglycerols and regulates the levels of triglycerides, fatty acids and cholesterol.
(c) Protein metabolism: removes excess amino acids from the bloodstream; these can be used to synthesise proteins or converted to lipids or glucose.
(d) Removal of drugs (inactivation) and hormones: deamination, converts toxic ammonia to fairly harmless urea.
(e) Excretion of bilirubin and synthesis of bile and bile salts: bile salt plays a role in lipid digestion (emulsification).
(f) Storage of vitamins: fat-soluble vitamins (A, B12, D, E, K) are absorbed by the blood and stored in the liver.
(g) Storage of minerals: converts iron reserves to ferritin and stores this protein-iron complex.
(h) Phagocytosis by Kupffer's cells in the sinusoids: engulf old or damaged RBC, cell debris and pathogens.
(i) Activation of vitamin D: it modifies cholecalciferol (provitamin D3) prior to its activation in the kidney.
(j) Storage of blood.

12. *What is the effect of pepsin on protein?*

Pepsin is a stomach enzyme that digests (hydrolyses) protein into smaller polypeptide pieces. As a pink/violet colour appears, this indicates that polypeptides are present – so pepsin must have converted some large protein molecules into smaller polypeptide units (the first stages of digestion have occurred).

13. *Consider two men: one of 70 kg with 15% BF and the other of 100 kg and 12% BF.*

   (a) Which one has a lower percentage of fat?
   (b) Calculate 12% of 100 kg and 15% of 70 kg. Which one has the greater mass of body fat?

   The 100 kg chap is leaner (has less % body fat).
   The 100 kg chap has a greater mass of fat (12 kg) than the little guy (who has 10.5 kg).

14. *In what clinical situations would skinfold measurements (or body mass index) provide useful information about the nutritional status of clients?*

   (a) Patients with morbid obesity – to look for changes in their fat levels (perhaps the callipers would not fit over the skinfold!).
   (b) Comatose patients who are being fed intravenously using a calculated kJ value diet – to see if energy intake is adequate.
   (c) Anorexia nervosa sufferers – to see if they are putting on weight.
   (d) Patients who are not very active (are not athletes) and who have low body fat – to check if they are undernourished (?).
   (e) Cancer patients who are not eating much (cancer cachexia).
   (f) To monitor cancer patients who may be experiencing cancer cachexia (reduced desire to eat).
   (g) To monitor problems of absorption from the gut.
   (h) To classify the heart disease risk category of someone who is overweight.
   (i) To classify the diabetic risk category for someone who is overweight.
   (j) Perhaps to monitor weight in patients prescribed diabetic drugs as some diabetes drugs cause weight gain.

Exercise 7: Endocrine System

1. *(a) What are hormones? (b) Describe their chemical classification (and give an example of each).*

   (a) Hormones are chemical messengers made in small quantities by endocrine cells and released into circulation. They travel through the bloodstream and bind to specific receptors in their "target cells" and produce a profound effect.
   (b) Hormones are classified into one of three structures: (1) amino acid derivative ("tyrosine" (e.g. adrenalin) or "tryptophan" (e.g. melatonin); (2) peptide derivative, which are those consisting of long chains of amino acids, e.g. prolactin (and ADH, OT, GH), or of long chains of AA with a carbohydrate side chain (also called glycoproteins), e.g. follicle-stimulating hormone (and TSH, LH); and (3) lipid derivatives, which are steroid hormones (e.g.

testosterone, estradiol, progesterone) and eicosanoids (e.g. leucotrenes, prostaglandins. These hormones are soluble in plasma membrane).

2. *Briefly describe the anatomical relationship between the hypothalamus and the pituitary gland.*

   The pituitary lies below (inferior to) the hypothalamus, connected to it by a stalk (the infundibulum) and cradled by a bony depression in the skull called the hypophyseal fossa within the sella turcica.

3. *What are the three ways that the hypothalamus controls the endocrine system and integrates the activities of the nervous and endocrine systems?*

   The hypothalamus provides the highest level of endocrine control in three ways:

   (a) The hypothalamus produces the hormones ADH and oxytocin, which are transported along axons to the posterior pituitary to be stored and released into the blood.

   (b) The hypothalamus secretes regulatory hormones that control endocrine cells in the anterior pituitary gland (releasing hormones, e.g. gonadotropin-releasing hormone (GnRH) and inhibitory hormones that stimulate the anterior pituitary gland to secrete hormones). In turn, the hormones released by the anterior pituitary control the activities of endocrine cells in the thyroid, cortex of adrenal glands and reproductive organs.

   (c) The hypothalamus contains "autonomic centres" that exert neural control (via impulses through nerve fibres) over the endocrine cells of the medullae of the adrenal glands. That is, when the "sympathetic division" of the autonomic nervous system is activated, adrenal medullae release hormones (epinephrine and norepinephrine) into the bloodstream.

4. *Briefly describe the structure of the pituitary gland and name the two hormones stored in the posterior pituitary.*

   The pituitary gland sits nestled within the sella turcica, a depression in the sphenoid bone. It is inferior to the hypothalamus and is connected to it by the infundibulum. The infundibulum lies between the optical chiasm and mammillary bodies. It is divided into the anterior lobe (glandular epithelial tissue), which is able to produce hormones (nine kinds of hormones), and a posterior lobe (made of neural tissue), which stores hormones that are made in the hypothalamus and transported to the posterior pituitary through the axons of nerve fibres. ADH and oxytocin are stored in the posterior pituitary.

5. *How are hormone receptors related to hormones?*

   Receptors are sites on the plasma membrane (extracellular receptor), organelles within the cell or in the nucleus (intracellular receptor) that have a shape that allows only that specific hormone with a complementary shape to bind to it. That is, receptors are specific to their hormone.

6. *How does the difference in the chemical structure of hormones determine where their receptors are located?*

   Hormones are classified into two groups: amino acid (AA)-based hormones (AA derivatives and peptide hormones) and lipid derivatives (steroid hormones and eicosanoids).
   The plasma membrane of a cell is made of lipid molecules; steroid hormones are lipid soluble, so they can diffuse through the cell membrane to bind to receptors in the cytoplasm or nucleus. Eicosanoids are lipid soluble, so they diffuse across the membrane to receptor proteins on the inside of the cell membrane.
   Hormones that are not lipid soluble (amino-acid-derived hormones, e.g. catecholamines) and peptide hormones are unable to penetrate the cell membrane. So their receptors are membrane proteins on the outside of the cell membrane, except for thyroid hormones, which can cross the membrane by diffusion or by a carrier mechanism and bind to receptors within the nucleus or mitochondria (energy dependent).

7. *Describe an example that shows how the hypothalamus is able to control the secretion of a hormone from a "subservient" endocrine gland.*

   The hypothalamus releases TRH to the anterior pituitary (in response to decreased T3 and T4 concentration in the blood or low body temperature, the anterior pituitary releases TSH).
   TSH causes thyroid follicles to release T3 and T4 into the blood (which increases their concentrations back to normal); whereupon, they move to their target cells.
   (Or) The hypothalamus releases GnRH, which stimulates the anterior pituitary to release FSH (and LH).
   FSH stimulates the testes to release inhibins and the ovaries to release inhibins and oestrogens.
   LH stimulates the testes to release androgens and the ovaries to release progesterones and oestrogens.

8. *Complete the table below: list the major (one or two) hormones produced; state a brief function for the hormone.*

Gland	Hormone(s)	Function
Thyroid	Thyroxin (T3,T4) Calcitonin	Raises the body's metabolic rate Lowers calcium concentration in the body fluid
Parathyroid	Parathyroid h.	Raises calcium concentration in the body fluid
Adrenal cortex	Aldosterone Cortisol (glucocorticoids)	Na+ to be re-absorbed in the kidneys in exchange for K+ Raises the rate of glucose synthesis and glycogen formation from fatty acids and proteins; has anti-inflammatory effects; inhibits the activities of the WBC

(continued)

Adrenal medulla	Adrenalin (epinephrine)	Speeds up the use of cellular energy and the mobilization of energy reserves; prepare for "action"
Pancreas	Insulin Glucagon	Lowers blood glucose Raises blood glucose
Testes	Testosterone	Allows the development and maintenance of masculine characteristics
Ovary	Estradiol Progesterone	Allows the development and maintenance of feminine characteristics Supports pregnancy
Pineal	Melatonin	(Suggested functions) inhibits reproductive functions, protects CNS neurons against free radicals, sets circadian rhythms
Anterior pituitary	Thyroid SH (TSH) ACTH Follicle SH Luteinising H Prolactin GH	Causes the secretion of thyroid hormones Causes glucocorticoid secretion, stimulates the release of steroid hormones by the adrenal cortex Causes oestrogen secretion, follicle development and sperm maturation Induces ovulation, the formation of corpus luteum, oestrogens and progesterone secretion Stimulates the production of milk Assists in growth, protein synthesis and lipid catabolism
Posterior pituitary	Oxytocin ADH	Promotes labour contractions and delivery, milk ejection, The re-absorption of water and an increase in blood volume and pressure
Thymus	Thymosins	Coordinates and regulates immune response

9. *Describe how amino-acid-based hormones exert their influence on target cells.*

   Proteins and peptides cannot penetrate cell membranes; AA-based hormones (first messengers) exert their effect via intracellular **second messengers** (e.g. cAMP) after binding to a receptor on the plasma membrane. The hormone bound to the receptor activates a "G-protein" inside the cell, which then moves along the membrane to activate an enzyme (adenylate cyclase) that produces cAMP. cAMP triggers a cascade of chemical reactions, producing increasing numbers of molecules at each step – this increase is called amplification!

Exercise 8: Urinary System

1. *Draw or trace a diagram of a kidney in the coronal section. Label the cortex, medulla, pyramids, columns, papilla, pelvis, major calyces and minor calyces.*
2. *Name the sections of the renal tubule in the order that filtrate passes through them.*

   Bowman's capsule, PCT, descending limb of the loop of Henle, ascending limb of the loop of Henle, DCT and collecting duct.

3. *What is meant by the terms glomerulus, vasa recta and Bowman's capsule?*

   Glomerulus: a capillary bed that filters blood cells and plasma proteins from the filtrate. It is fed by the afferent arteriole and drained by the efferent arteriole.
   Vasa recta: a "counter-current" capillary bed that surrounds the loop of Henle of juxtamedullary nephrons. It redistributes water and solutes that are re-absorbed in the medulla and stabilises the concentration gradient of the medulla. The vasa reacta returns water and solutes to the general circulation.
   Bowman's capsule: a structure that surrounds the glomerulus and receives the filtrate.

4. *With reference to the kidney tubule, distinguish between filtration and osmosis.*

   Filtration: the process of separating large structures in the blood (cells and plasma proteins) from smaller structures. It is driven by a pressure difference. It happens in the glomerulus.
   Osmosis: the process of movement of water through a semi-permeable membrane, from a solution with a high concentration of solutes to a solution with a lower concentration of solutes. It is a diffusion process driven by differences in solution concentration. It happens in the renal tubule, where water is recovered from the filtrate after Na ions are actively transported.

5. *Which parts of the renal tubule are impermeable to water?*

   The thick portion of the ascending limb of the loop of Henle but the collecting duct and proximal convoluted tubule in the absence of the antidiuretic hormone.

6. *Name and briefly describe the four stages of urine formation.*

   Filtration: blood cells and proteins are separated from (filtered out of) the rest of the blood and retained in the glomerular capillaries. Small particles (water, electrolytes, glucose, nitrogenous wastes etc.) pass into the Bowman's capsule and are known as the filtrate.
   Re-absorption: useful substances (water, electrolytes, glucose etc.) are reclaimed from the filtrate and returned to the blood.
   Secretion: some substances that were not present in the filtrate (H+, K+, HCO_3^- ions) are secreted through the tubule walls into the filtrate. Some substances that were re-absorbed from the filtrate are secreted back into the filtrate (urea, uric acid).
   Concentration of urine: the concentration of the filtrate is adjusted in the DCT and collecting ducts by the influence of aldosterone (which causes Na+ to be re-absorbed) and ADH (which causes water to be re-absorbed) depending on the body's state of hydration and the blood's osmolality. After concentration, the filtrate is known as urine.

7. *Describe the function of the juxtaglomerular apparatus.*

   The JG apparatus is the junction of the afferent arteriole, the efferent arteriole and the distal end of the ascending limb of the loop of Henle. It consists of the granular cells (JG cells) in the arteriole walls (which produce and release renin) and the macula densa cells in the tubule walls (which sense the concentration of Na+ and Cl^- ions in the filtrate). The macula densa causes the afferent arteriole to dilate if the concentration of Na+ and Cl^- ions in the filtrate is too low. Dilation increases the GFR, which causes more filtrate to be produced so fewer ions can be re-absorbed from the greater volume flowing through the tubule. The JG cells are mechanoreceptors and produce and release renin. They sense the degree of stretch (i.e. the blood pressure) in afferent arterioles. Reduced stretch ⇒ decrease in BP ⇒ release of renin ⇒ formation of angiotensin II, which causes the constriction of arterioles throughout the body to raise BP. Angiotensin II also stimulates the release of aldosterone, which causes sodium ions to be re-absorbed from the DCT and collecting ducts. Water will "follow" sodium (through osmosis) back into the bloodstream to increase blood volume and so raise BP. Angiotensin II also stimulates thirst and increases ADH secretion.

8. *Describe the changes in the composition of the filtrate as it passes through the descending limb and then the ascending limb of the Loop of Henle.*

   In the descending limb (which is permeable to water but not to sodium or chloride) water leaves the filtrate by osmosis, so the volume of the filtrate decreases but the concentration of the filtrate increases. In the ascending limb (which is permeable to sodium and chloride but not to water), sodium and chloride are re-absorbed – but not water – so the concentration of urine decreases but the volume stays about the same.

9. *Explain how the kidney can produce concentrated urine when the body needs to.*

   The wall of the collecting duct is impermeable to water unless the hormone ADH is present. In this case, water is able to pass out of the collecting duct into the cells of the tubule wall. This happens because (1) the walls are permeable due to the ADH-regulated expression of aquaporins and (2) there is an osmotic gradient allowing water to move osmotically from the filtrate into the tubule cells. The osmotic gradient exists because of the high concentration of urea in the medulla of the kidney.

10. *Explain the effect of aldosterone on the composition of urine.*

    Maintaining sodium ion balance is aldosterone's primary role. It targets (primarily) the DCT, where it stimulates the re-absorption of sodium ions from the filtrate (and chloride ion is co-transported). As sodium is absorbed from the filtrate, potassium is secreted into the filtrate in a one-for-one exchange. If the DCT is permeable to water, then water will also be absorbed by the DCT. If a person is well hydrated and if aldosterone is present in the absence of ADH, large volumes of sodium-free urine can be produced.

11. *What is the role of ADH?*

    The antidiuretic hormone (ADH) makes the collecting duct (and distal convoluted tubule) of the nephron more permeable to water (i.e. pores for water to pass through (aquaporins) are inserted into the plasma membranes of the cells that make up the collecting duct). In this way, water can be re-absorbed from the filtrate (by osmosis), and the kidney will produce a more concentrated urine.

12. *What is the role of aldosterone?*

    Aldosterone will cause Na+ to be absorbed from the filtrate (and K+ to be secreted into the filtrate in exchange – this maintains electrical neutrality). This absorption increases the solution concentration in the cell cytoplasm, interstitial fluid and blood so that water will move (if ADH is present) from the filtrate into the cells etc. Hence, it could be said that aldosterone also causes more re-absorption of water.

13. *Describe the feedback response that results in:*

    (a) *ADH secretion*
    (b) *Aldosterone secretion*

    If the osmolarity of blood increases towards 300 mosmol/L, osmoreceptors in the hypothalamus include neurons that produce ADH notice (or ECF volume decreases or angiotensin II is present), and then ADH (= vasopressin) is secreted (the neurons release ADH in the posterior pituitary) to cause more water to be absorbed from the filtrate (the kidney conserves water). This conservation of water will limit the further increase in blood osmolarity (thirst will also be stimulated; that is, drinking water will reduce blood osmolarity).
    If Na concentration in the blood falls or blood volume or pressure decreases, the hormone rennin is released (then angiotensinogen is activated to angiotensin I, then to angiotensin II, and then aldosterone is released from the adrenal glands), aldosterone causes more Na+ to be absorbed by distal convoluted tubules and collecting ducts (by inserting more Na channels in the plasma membrane). The absorption of Na causes the [Na] of blood to increase, ECF volume to increase and BP to increase, so the stimulus to produce more aldosterone is removed.

14. *What is the function of each of the four structures of the nephron in urine formation?*

    1. Glomerulus: filtration of blood
    2. Proximal convoluted tubule: reabsorption of most of the required substances from the filtrate.
    3. Loop of Henle: absorption of water (from descending limb), absorption of Na (from ascending limb)
    4. Distal convoluted tubule: absorption of more Na (if aldosterone present) absorption of more water (if ADH is present)

15. *Write a paragraph describing the blood supply to the kidney.*

    I want the paragraph to contain: renal artery comes off the abdominal aorta.

I want afferent arteriole and efferent arteriole mentioned.
I want the peri-tubular capillaries and vasa recta to also be included.

16. *(a) How (and under what conditions) does the kidney excrete bicarbonate ions?*
(b) How (and under what conditions) does the kidney re-absorb or manufacture new bicarbonate ions?

(a) When the body is in alkalosis, type-B intercalated cells in collecting ducts effectively secrete bicarbonate ions in the reverse of the process described in part b) below.
(b) The body excretes HCO3$^-$ into filtrate all the time because the body usually has excess acid to excrete but particularly during acidosis.

In kidney tubule cells, CO_2 combines with water to form carbonic acid. Carbonic acid splits into H^+ and HCO_3^- (i.e. new bicarbonate is formed). HCO_3^- is passed to the peritubular capillary (absorbed by the body) – in exchange for Cl^- or Na^+.
H^+ is secreted into the filtrate and combines with HCO_3^- there (which was filtered from the blood) to form carbonic acid – which the disassociates into CO_2 and water. CO_2 is re-absorbed to make more carbonic acid.
Amino acid glutamine is deaminated to form NH4$^+$ (or NH3) and HCO_3^- – ammonia is secreted into the filtrate, and bicarbonate is absorbed into the blood.

Exercise 9: Nervous System

1. *Distinguish between (a) the CNS and the PNS, (b) the efferent nervous system and the afferent nervous system, (c) the autonomic NS and the somatic NS, (d) the sympathetic NS and the parasympathetic NS.*

(a) CNS = the brain and the spinal cord. It integrates, processes and co-ordinates sensory data and motor commands. The brain is the seat of higher functions. PNS = neural tissue outside the CNS. It consists of motor nerves, sensory nerves, receptors and autonomic nervous system nerves. The PNS delivers sensory information to the CNS and carries motor commands from the CNS to peripheral tissues and systems.
(b) The efferent system carries nerve impulses away from the CNS and towards some effector, e.g. a muscle (including skeletal, smooth and cardiac muscles) or gland; afferent system fibres carry impulses towards the CNS from peripheral sensory receptors.
(c) The somatic NS sends motor messages from the CNS to the skeletal muscle and is under our conscious control. The autonomic NS conducts impulses from the CNS to the cardiac muscle, smooth muscle and glands (i.e. our viscera). It is not under our voluntary control.
(d) The sympathetic division stirs you up and mobilises the body during emergency (stressful) situations (e.g. increases HR, dilates bronchioles, redirects blood supply, decreases gut motility, stimulates sweat glands). The parasympathetic division promotes non-emergency function and conserves

energy (e.g. decreases HR, constricts bronchioles and pupil, increases gut motility and secretions).

2. *Describe the role of the sympathetic division of the ANS and its effects on the body.*

   When stimulated, sympathetic preganglionic neurons release Ach at synapses with ganglionic neurons, and most sympathetic ganglionic neurons release NE at their varicosities, and the NE released affects its targets until it is re-absorbed or inactivated by enzymes; for example, it stimulates the release of renin from the kidney. Renin causes the formation of angiotensin II, which stimulates the release of aldosterone (from the adrenal cortex), ultimately raising blood pressure. Sympathetic preganglionic fibres also enter the centre of the adrenal gland (adrenal medullae) and stimulate the release of epinephrine (adrenaline) and norepinephrine. The bloodstream carries NE and E throughout the body, where they cause changes in the metabolic activities of many different cells. For example, NE and E increase metabolic rate and BP and engage in a "fight or flight" response.

3. *Draw a simple diagram of CNS meninges consisting of three concentric circles. Label the three meninges, the epidural space, the sub-arachnoid space and the CSF. Further draw two needles: one delivering an epidural anaesthetic and the other in place for a lumbar puncture to draw out CSF.*

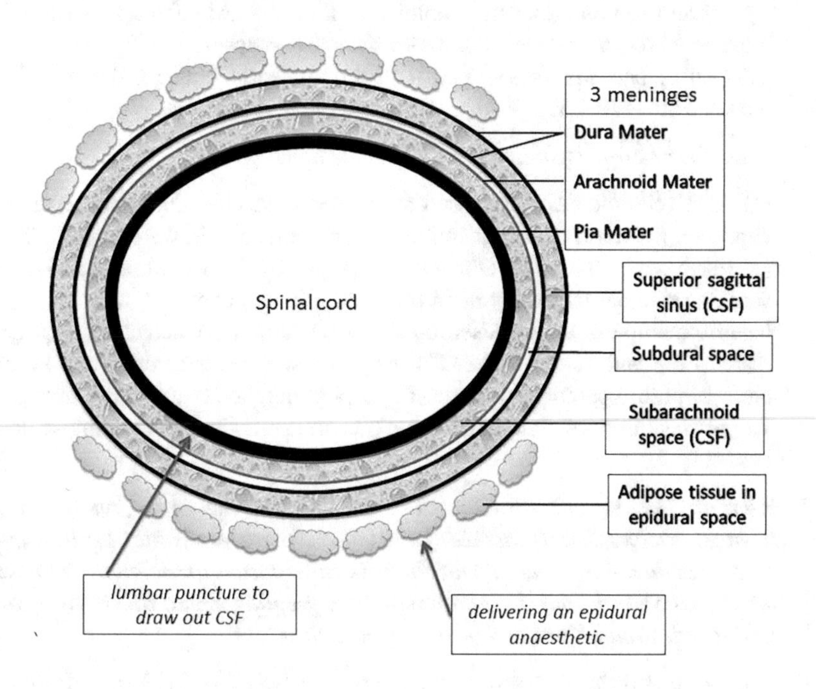

(Lumbar puncture draws CSF from within the sub-arachnoid space. An epidural anaesthetic is delivered into the epidural space, superficial to the dura mater but inside the vertebral foramen.)

4. *Distinguish between a unipolar and a multipolar neuron.*

   A "multipolar" neurone has more than two processes extending from the cell body: one of them is an axon, and the rest are dendrites. A "unipolar" neurone has only one process extending from the cell body. The dendrites and axons are continuous, and the cell body lies off to one side.

5. *(a) What is a "graded potential"?*

   A typical stimulus produces a temporary localised change in the resting membrane potential. The effect, which decreases with distance from the stimulus, is called a graded potential.

   (b) What is an "action potential"?

   If the graded potential is large enough, it triggers an action potential in the membrane of the axon.
   "Na gates" open, and extracellular Na rushes into the axon till the gate closes. Then "K gates" open, and intracellular K rushes out of the axon till the gate closes.
   An "action potential" is an electrical impulse that is propagated along the surface of an axon and does not diminish as it moves away from its source. This impulse travels along the axon to one or more synapses.
   The resting potential is −90 mV. An action potential changes this to +30 mV then back to −70 mV.

6. *Name four neurotransmitters. What is their function?*

   (1) Acetylcholine (Ach) released at the neuromuscular junction; (2) norepinephrine, the main neurotransmitter of ganglion cells in the sympathetic NS; (3) nitric oxide (NO); (4) dopamine; (5) and GABA (gamma amino butyric acid), a principal inhibitory neurotransmitter in the brain.
   Neurotransmitters are released by the axon terminal when a nerve impulse reaches the end of the nerve. The neurotransmitters then cross the synaptic cleft, bind to receptors on the post-synaptic membrane and cause changes in the permeability of that membrane (i.e. transmit the nerve impulse to the next cell)

7. *What are the five main structures of the CNS? In which of the five are the following structures/areas located: conus medullaris, respiratory centres, motor areas, cauda equina, autonomic control centre, visual association areas, area for the control of body temperature, cervical enlargement, arbor vitae, basal nuclei, substantia nigra and posterior median sulcus.*

   The brain and the spinal cord make up the CNS. The CNS includes the (1) cerebrum, (2) diencephalon, (3) brain stem, (4) cerebellum, and (5) spinal cord.

Cerebrum: motor areas, visual association areas and basal nuclei
Brain stem: substantia nigra (midbrain), respiratory centres (medulla oblongata) and autonomic control centre (the pons and medulla oblongata)
Diencephalon: area for the control of body temperature
Cerebellum: arbor vitae
Spinal cord: conus medullaris, cervical enlargement, cauda equina and posterior median sulcus

8. *What constitutes a reflex arc? Describe the stretch reflex.*

The route followed by nerve impulses to produce a reflex is called a reflex arc. It is a rapid predictable motor response (unlearned and involuntary) to a stimulus (without processing by the brain).
The reflex arc involves five steps:

(a) The arrival of a stimulus and the activation of a receptor.
(b) The activation of a sensory neuron (transmits afferent impulses to the CNS).
(c) Information processing in the CNS (sensation relayed to the brain by axon collaterals).
(d) The activation of a motor neuron (conducts efferent impulses to an effector organ).
(e) The response of a peripheral effector (muscle fibre or gland cell).

Stretch reflex: this is an example of a monosynaptic reflex because there is only one synapse. The patellar reflex is an example of a stretch reflex. The stimulus is a tap on the patellar tendon. This stretches receptors within the quadriceps muscles. The response is a brief contraction of those muscles, which produces a noticeable kick.

9. *What chemical is stored in the synaptic vesicles?*

A neurotransmitter.

10. *Distinguish between a unipolar, a bipolar and a multipolar neurone.*

Uni = one fused axon dendrite (one process attached to cell body).
Bi = one axon and one dendrite attached to each cell body (i.e. two processes).
Multi = many dendrites attached to the cell body (and one axon).

11. *Draw a cross-sectional diagram of the spinal cord and label it with the structures: spinal meninges, grey matter, white matter, dura mater, arachnoid, pia mater, dorsal (posterior), grey horns, ventral (anterior) grey horns, sensory fibres, dorsal (posterior) root ganglion, spinal nerves, grey commissure, lateral grey horns, central canal, anterior median fissure and posterior median sulcus.* (Consult your textbook.)

12. *What is the cauda equina?*

(Horse's tail) the spinal cord ends at about L1. Inferior to this, the spinal nerves continue within the vertebral foramina (before they exit). This aggregation of individual fibres has the appearance of separate "hairs", like the tail of a horse.

13. *What is the difference between the spinal cord at L1 and L3?*

    The spinal cord covered by the pia, arachnoid and dura maters ends as a tapering structure called the conus medullaris at about L1. The pia mater continues as a fibrous extension (the filum terminale) to the posterior coccyx. Below L2, spinal nerves are outside the pia but enclosed by the arachnoid and dura and are in contact with CSF. The spinal cord can be punctured by an epidural needle that is advanced too far, but this potential danger is avoided by administering below L2.

14. *Which two meninges enclose the CSF?*

    Arachnoid mater and pia mater.

15. *What is the name of the "space" that contains CSF?*

    The sub-arachnoid space.

16. *If a patient has raised intracranial pressure, would CSF pressure in the lumbar region (measured by lumbar tap) be higher or lower than normal?*

    Higher! (Pascal's principle = increased pressure applied at one place (in the cranium) is transmitted throughout the enclosed liquid (CSF)).

17. *What structures separate the epidural space from the sub-arachnoid space?*

    Dura mater and arachnoid mater.

18. *Does the epidural space contain major nerves?*

    YES. It contains adipose tissue between the dura mater of the spinal cord and the bone of the vertebrae, but also the spinal nerves traverse this space – hence, they can be anaesthetised.

19. *Describe where the cell bodies of sensory neurons are located.*

    The dorsal root ganglion, i.e. outside the CNS.

20. *Describe where the cell bodies of motor neurons are located.*

    The ventral horn (of grey matter in the spinal cord), i.e. within the CNS.

Exercise 10: Special Senses

1. *Describe the function of the (a) cornea, (b) lens, (c) retina, (d) canal of Schlemm, (e) choroid and (f) pupil.*

    The cornea converges (refracts) and focuses light rays.
    The lens focuses light from objects onto the retina (accommodates).

The retina absorbs light photons (in rods and cones) and converts energy into nerve impulses (in bipolar and ganglion cells, which leave the eye via the optic nerve).
The canal of Schlemm allows the aqueous humour in front of the anterior compartment of the eye to drain away – its blockage can cause raised intraocular pressure damage to the retina.
The choroid is sandwiched between the retina and sclera. It contains the blood capillaries that supply the retina.
The pupil is the transparent space (in the centre of the iris) that allows light to enter the eye. It may expand and contract.

2. *Describe the role of the ciliary body and suspensory ligaments in the process known as accommodation.*

   The ciliary body is a muscle that is shaped like an annulus. When it contracts, its central hole gets smaller in diameter, the tension on the attached suspensory ligaments of the lens is relaxed and the elastic lens can ooze back towards a spherical shape (to focus light from near objects). When the ciliary body relaxes, its hole enlarges; this pulls on the suspensory ligaments, which stretch the lens into a flatter shape and enable the lens to focus light from distant objects.

3. *Why does the retina have a blind spot?*

   The axons and neural tissue lie above the rods and cones; in order for the axons to carry impulses to the brain, they come together into the optic nerve, which exits the posterior chamber at the blind spot. At this spot, there are no rods and cones because the optic nerve has blasted through where they would otherwise have been. There is no light sensitivity at this spot.

4. *Describe the effect on the light rays of each type of lens (diverge/converge).*

   A concave lens diverges light rays (so with corrective eyeglasses, the focal spot is moved further from the eye lens).
   A convex lens diverges the light rays (so the focal spot is moved closer to the eye lens).

5. *What are the names commonly given to these two types of lenses?*

   Concave and convex lenses.

6. *Which one produces a magnified image when placed over some writing on a page?*

   Convex.

7. *Which one has the same curvature as your eye's lens?*

   Convex.

8. *Define what is meant by hyperopia.*

   (Long-sightedness = lens-to-retina distance is too short.) The eye can accommodate seeing distant objects clearly, but close objects are out of focus as the light from them is focused "behind" the retina – a convex lens placed in front of the eye provides the extra convergence for clear vision.

9. *Describe the function of the (a) pinna, (b) ossicles, (c) Eustachian tube and (d) cochlea.*

   Pinna funnels sound onto the tympanic membrane.
   Ossicles in the middle ear convert/transfer a vibration in the air (detected by a tympanic membrane) to a vibration in the cochlear liquid (they are an impedance-matching device that also provides a mechanical advantage, i.e. some amplification).
   The Eustachian tube (normally squashed flat) allows the air space of the middle air to be vented, i.e. allows air pressure in the middle ear to be equalised to that outside.

10. *What are the average sound frequency endpoints of the audible range for humans?*

    It is ~40 Hz to 18,000 Hz for youngsters. Mature people will not hear beyond 15,000 Hz.

Exercise 11: Blood

1. *Outline the functions of the blood and state which components (plasma, erythrocytes, leucocytes and platelets) are responsible for which properties of blood.*

   Functions:

   (a) Transporting dissolved gases, nutrients, hormones and metabolic wastes: blood carries oxygen from the lungs to the cells and CO_2 from the cells to the lungs, distributes nutrients absorbed by the digestive tract or released from storage in adipose tissue or the liver, carries hormones from the endocrine system to target cells and receives wastes from cells and carries them to the kidney for excretion.

   (b) Regulating the pH, temperature and ion composition of interstitial fluids: blood absorbs and neutralises acids generated by active tissues; it absorbs the heat generated by active skeletal muscles and redistributes it to other tissues or heat is lost from the body across the surface of the skin.

   (c) Defending against toxins and pathogens: blood transports WBC and specialised cells to fight infections or remove debris; it delivers antibodies and proteins to attack invading organisms.

(d) Restricting fluid losses at injury sites: blood contains enzymes and other substances that respond to breaks in vessel walls by initiating the process of clotting, coagulation and clotting act as a temporary patch that prevents further blood loss.

Plasma makes 55% of the volume of whole blood; it contains water, ions, clotting factors, nutrients, albumins, globulins and fibrinogen. The function of plasma includes helping maintain blood volume and osmotic pressure (albumins), defence (globulins), clotting (fibrinogen) and transporting lipids and small ions (globulins).
Erythrocytes transport respiratory gases. Each developing RBC loses cellular organelles. Molecules of haemoglobin make up more than 95% of its intercellular proteins. Of the oxygen carried by blood, 98.5% are bound to Hb molecules inside RBCs.
Leucocytes contribute to the body's defences; they help defend the body against invasion by pathogens. They also remove toxins, wastes and abnormal or damaged cells. Leucocytes comprise the non-specific defence (neutrophils, eosinophils, basophils and monocytes) and specific defence (lymphocytes) of the body.
Platelets are involved in the blood clotting process. They release chemicals important to the clotting process, form a temporary patch in the wall of damaged blood vessels and reduce the size of a break in the vessel wall.

2. *What is the structure, composition and function of RBC and outline briefly the functions of white cells.*

   RBC structure: each RBC is a bi-concave disc about 7 μm in diameter with a thin central region and a thicker outer margin.
   Composition: molecules of haemoglobin (Hb) make up more than 95% of its intracellular proteins. They have no nucleus or mitochondria.
   Function of RBC: they are involved in the transport of respiratory gases; 98.5% of the oxygen carried by the blood is bound to Hb molecules inside RBCs.
   Function of WBC: WBC contribute to the body's defences; they help defend the body against invasion by pathogens. Lymphocytes (T, B and NK cells) recognise specific "non-self" antigens. Neutrophils and eosinophils are microphages. Basophils promote inflammation. Monocytes are macrophages, and they engulf foreign bodies and digest them. They also remove wastes and abnormal or damaged cells.

3. *State briefly the effects of iron deficiency, amino acid deficiency and vitamin B12 deficiency in the blood.*

   Iron is required to make haemoglobin for RBC. Iron deficiency will affect RBC formation because Hb has four chains of polypeptides and each chain contains a single molecule of heme and each heme holds an iron ion to interact with an oxygen molecule, forming oxyhaemoglobin (i.e. four atoms of iron per Hb). A deficiency in iron will result in anaemia (low levels of RBC) and is called iron-deficiency anaemia.

AA deficiency will affect the synthesis of plasma proteins: plasma proteins (e.g. transferrin) serve as carriers for other molecules. Many types of small molecules bind to specific plasma proteins and are transported from organs that absorb these proteins to other tissues for utilisation. Proteins also act as a pH buffer to keep blood at a stable pH. Plasma proteins (fibrinogen, prothrombin and plasminogen) interact in specific ways to cause the blood to coagulate. Plasma proteins govern the distribution of water between the blood and tissue fluid by producing what is known as a colloid osmotic pressure. Protein deficiency anaemia is an anaemia that results from an inadequate intake of dietary protein that disrupts normal haemoglobin synthesis.
Vitamin B12 deficiency impairs RBC production, causing megaloblastic anaemia (= red blood cells become larger than normal)

4. *Outline the role of platelets in haemostasis and describe the three main processes in coagulation, including the differences between extrinsic and intrinsic formation of prothrombin activator (exclude clotting factor names).*

   Platelets' role in haemostasis: (1) releasing chemicals important to the clotting process; they help initiate and control the clotting process; (2) forming a temporary patch in the walls of damaged blood vessels; the platelet plug can slow blood loss while clotting takes places; and (3) reducing the size of a break in the vessel wall after the clot has formed. Three main processes in coagulation: (1) the extrinsic pathway – it begins when damaged endothelial cells or peripheral tissues (i.e. external to the blood) release factor III, which initiates the formation of prothrombinase, which converts prothrombin to thrombin; (2) the intrinsic pathway – it is more complex and occurs more slowly; damaged tissues cause platelet damage, and the platelets that are internal to the blood (i.e. intrinsic pathway) release phospholipids, which combine with various clotting factors to produce prothrombinase; and (3) the common pathway: it begins when prothrombin 9 sometimes called factor X complexes with calcium phospholipid and modified factor V to convert prothrombin to thrombin. Thrombin then completes the clotting process by converting fibrinogen to insoluble fibrin.

5. *State the role of vitamin K in clotting and explain how uptake from the gut can be affected by disorders in fat absorption (fat soluble enzyme).*

   Vitamin K affects the blood clotting process; adequate amounts of vitamin K must be present for the liver to synthesise four of the clotting factors, including prothrombin. Vitamin K is a fat-soluble vitamin that is absorbed with dietary lipids. We obtain half of our daily requirement from the diet. Hence, if there is a disorder in fat absorption, vitamin K absorption will be affected. Vitamin K deficiency will cause the eventual breakdown of the common pathway due to a lack of clotting factors.

6. *Explain the clinical uses of heparin and warfarin in relation to their speed of effectiveness as anticoagulants.*

   Heparin accelerates the activation of antithrombin-III and prevents the conversion of prothrombin to thrombin. Clinically, it used extensively post-operatively to impede or prevent clotting in stroke victims. It must be administered intravenously. It is rapid acting.
   Warfarin is antagonist to vitamin K (it prevents the formation of vitamin-K-dependent clotting factors). Warfarin can be taken orally and is slower acting, taking days to have a therapeutic effect. A standard anticoagulation regime would be immediate IV heparin administration continued while warfarin builds up to therapeutic levels, allowing the patient to eventually go home on oral warfarin alone.

7. *Explain transfusion reactions by your knowledge of the ABO blood group and relate newborn haemolytic disease to the Rh blood group.*

   The **ABO group** is based on the presence or absence on RBC of two antigens called type A and type B.
   Blood group A only has type A surface antigen on RBC and anti-B agglutinins in the plasma.
   Blood group B only has type B surface antigen on RBC and anti-A agglutinins in the plasma.
   Blood group AB has both type A and type B surface antigens on RBC but does not contain anti-A or anti-B agglutinins in the plasma.
   Blood group O has neither surface antigen on RBC and contains both anti-A and anti-B agglutinins in the plasma.
   Hence, these agglutinins will attack the antigens on foreign RBC. When these antibodies attack, the foreign cells agglutinate (clump together), causing agglutination and haemolysis of the affected RBC.
   Haemolytic disease of the newborn (HDN): HDN is an RBC-related disorder caused by a cross-reaction between the foetal and maternal blood types. If the mother is Rh− and the first baby is Rh+, during delivery, foetal blood may enter the maternal system from the placenta; hence, the mother will produce anti-Rh agglutinins. It will not affect the first baby; however, these agglutinins can cross the placenta in a subsequent pregnancy, and if the second baby is Rh+, the agglutinins will attack the RBC of the second baby, causing HDN.

8. *What are whole blood, packed cells and cryoprecipitate (clotting factors) used for in blood transfusions?*

   Whole blood: the combination of plasma and the formed elements; used to increase blood volume (e.g. after haemorrhage).
   Packed cells: boost haemoglobin to restore oxygen carrying capacity without placing hypervolaemic stress on the system.
   Plasma: RBCs are returned to the donor, the fluid matrix of blood. It is used to make 17 products for various trauma burns, cancer and blood disease treat-

ments. **Cryoprecipitate** contains blood clotting substances, called fibrinogen and factor VIII. Cryoprecipitate can be used for patients with particular deficiencies of these proteins. Cryoprecipitate is also used for trauma and liver transplants.

9. *One microlitre (1μl) of blood contains about 5 million (5×10^6) RBC; each RBC contains about 280 million (280×10^6) Hb molecules. Estimate how many Hb molecules are there in an adult human.*

   An adult female contains 4–5 l of whole blood. An Adult man contains 5–6 l of whole blood

$$\left(280\times10^{6}\right)\times\left(5\times10^{6}\right)\times5\times10^{6}\ \mu l=7\times10^{21}\ \text{Hb molecules.}$$

10. *Name the three types of white blood cells depicted below.*

Neutrophil

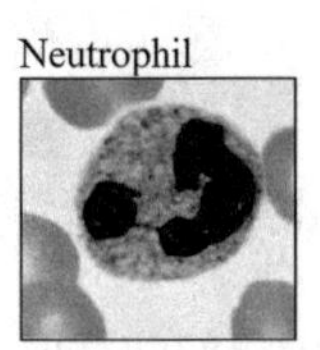

Lymphocyte

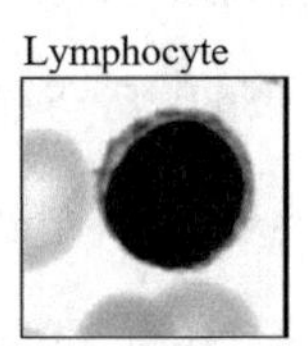

Monocyte

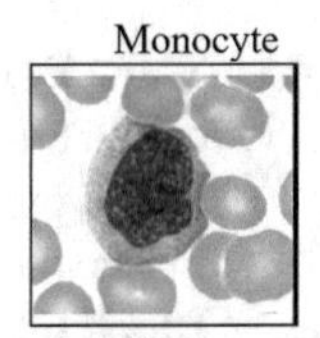

Exercise 12: Cardiovascular System: Anatomy of the Heart

1. *Which heart valve is also known as the tricuspid? (Which other heart valves also have three flaps?)*

   The right atrioventricular valve (the aortic and pulmonary valves).

2. *Which of (ventricular) systole or diastole refers to the contraction of the heart muscle?*

   Systole.

3. *Apart from the heart, what other structures lie within the mediastinum?*

   Thymus, oesophagus, the great vessels of the heart (e.g. descending aorta, ascending vena cava), trachea (in the upper part), the phrenic nerve, the cardiac nerve, the thoracic duct and the lymph nodes of the central chest.

4. *Arrange (write down) the four chambers of the heart (LA, RA, LV, RV), the four valves (R a-v, L a-v, Pul v, Ao v) and the blood vessels (aorta, pulmonary trunk, pulmonary veins, vena cava) in the order in which blood flows through them.*

   (Superior and inferior) vena cava, RA, Ra-v valve (tricuspid), RV, pulmonary valve, pulmonary trunk (right and left pulmonary arteries), (lungs) pulmonary veins, LA, La-v valve (= mitral, bicuspid), LV, aortic valve and aorta (systemic circulation).

5. *Name the membrane that adheres to the outside of the myocardium.*

   Visceral pericardium (epicardium).

6. *Distinguish cardiac muscle from skeletal muscle.*

   Cardiac muscle: the cardiac muscle is involuntary; cardiac muscle cells are relatively small and have a single centrally placed nucleus, although a few have two or more nuclei; the cells are generally branched and are interconnected by "intercalated discs".
   Skeletal muscle: the skeletal muscle is voluntary; skeletal muscle cells are cylindrical and much longer than cardiac muscle cells, and they have multiple nuclei; cells are not branched, and there are no intercalated discs presented in skeletal muscles. They are adjacent muscle fibres tied together by connective tissue fibres.

7. *Where are coronary arteries, and what do they do?*

   The coronary arteries lie over the myocardium and supply the muscle tissue of the heart with oxygenated blood. The two openings to the coronaries (left and right coronary arteries) lie just superior to the aortic valve so that at diastole, when the aortic valve closes, blood can flow into the openings to the left and right coronary arteries.

8. *What do the heart valves do as the ventricles progress through systole?*

   At the commencement of systole, the atrioventricular valves shut (the chordae tendineae and papillary muscles prevent the valve flaps from being pushed up into the atria); this traps blood in the ventricles so that the pressure in the blood contained in the ventricles rises as the ventricles apply compression. When the pressure of blood in the ventricles exceeds the back pressure exerted on the aortic valve (or the pulmonary valve for the right ventricle) by the blood already in the aorta (or pulmonary trunk), the ventricular blood forces the aortic valve (or the pulmonary valve) open and pushes the aortic blood further along the aorta and the ventricular blood into the aorta (or pulmonary arteries).

9. *Why is the myocardium of the left ventricle so thick?*

   The LV has to pump blood through the systemic circuit. The length of blood vessels in this circuit causes it to have a high resistance to blood flow; this means that the LV has to pump hard (harder than the RV) in order to create sufficient pressure to ensure that blood flows throughout the systemic arteries and arterioles. The RV, on the other hand, does not need to work very hard to push blood through the pulmonary circuit because the lungs are close to the heart, and the pulmonary blood vessels are relatively short and wide.

10. *On the schematic (i.e. not anatomically faithful) diagram of the heart below, draw in and label the aortic and pulmonary valves and label all the structures listed in question 4. In addition, use arrows to indicate the flow of blood.*

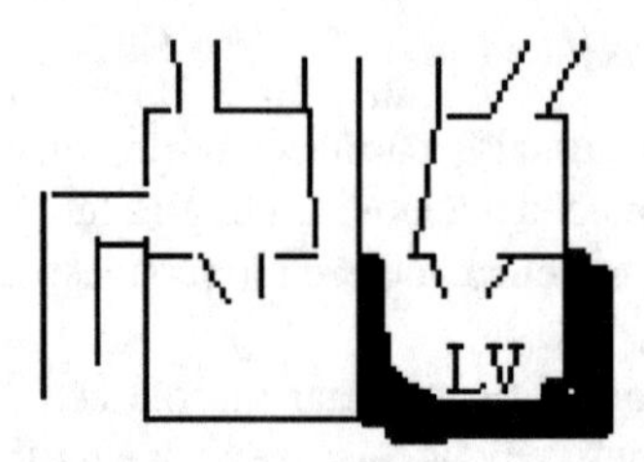

11. *What is the general function of the heart valves, and what would happen to the circulation of blood through the heart if the action of the valves was impaired?*

    Valves allow blood to flow in one direction only (the correct way). The valves close to prevent backflow. If valve function is impaired, some blood would flow the wrong way (backwards); this would result in a "murmur" and would make the heart pump less efficiently.
    (Heart valves may become impaired for a variety of reasons. Rheumatic fever, the most common cause of valve malfunction, causes a valve to stiffen over time. This limits the ability of the valve to open and close properly. Some people are born with heart valve abnormalities, which may be corrected at birth or later in life. Other people may acquire valve damage from infection (as with bacterial endocarditis) or other diseases. The results are a tight, rigid valve limiting forward blood flow (called a stenotic valve) or a valve that does not close properly, permitting backflow (called a regurgitant valve).

12. *How does the structure of the left ventricle differ from that of the right?*

    The LV has a much thicker muscle wall than the RV. The LV also is circular in cross-section, while the RV is a "crescent moon" shaped pocket attached to the LV.

13. *How do these differences relate to resistance to blood flow in pulmonary and systemic circulations?*

    Resistance to blood flow in a systemic circuit is higher than in a pulmonary circuit, so high blood pressure is required to overcome the resistance to flow (and so push blood into the arterial circulation); thus, a strong muscle (= thick) is required to provide this higher BP. Pulmonary resistance to blood flow is lower (because the length of blood vessels in the pulmonary circuit is shorter than in the systemic circulation. So a less powerful pump is required.

14. *Describe how and during which part of the cardiac cycle blood enters the coronary arteries.*

    Openings to the two coronary arteries are in the wall of the aorta, immediately beyond the tricuspid aortic valve. During heart contraction (systole), the aortic valve is open and the coronary artery openings are squashed shut.
    During diastole, the aortic valve shuts as blood fills the cusps when blood from the heart stops rushing into the aorta.
    Thus, during diastole, the coronary arteries are no longer squashed shut, and blood from the aorta then enters the coronary arteries.

15. *What effect would a blockage of the coronary arteries have on the myocardium (heart muscle)?*

    (Part of) the myocardium would not be sufficiently perfused (not receiving oxygen); this means it would be ischemic (i.e. lacks oxygen), or some tissue would die (an infarction). That is, a "heart attack" (= MI) ensues. Rapid medical intervention will minimise the size of the infarct. The MI may or may not be fatal.
16. *How many cusps does the RIGHT atrioventricular valve have? What is the other name for this valve?*

    Three, so it is called the tricuspid valve (or right parachute valve).
17. *How many cusps does the LEFT AV valve have? What is the other name for this valve?*

    Two. It is also known as the mitral valve.

Exercise 13: Blood Vessels

1. *What are the three tunics that comprise the walls of veins and arteries?*

   Tunica intima (or tunica interna), tunica media and tunica externa (or tunica adventitia).
2. *Name some differences between veins and arteries.*

   (a) Vessel walls: veins have thinner walls than arteries; the tunica media of an artery contains more smooth muscle and elastic fibres than does that of a vein; the tunica media of an artery also has an internal and an external elastic membrane, whereas the tunica media of a vein does not.
   (b) Vessel lumen: when not opposed by blood pressure, the elastic fibres in the arterial walls recoil, constricting the lumen. Veins are larger in diameter than the corresponding arteries. Arteries usually keep their cylindrical shape, but veins tend to collapse.Veins typically contain valves, internal structures that prevent the backflow of blood towards the capillaries.
   (c) Veins carry blood towards the heart, while arteries carry blood away.
   (d) Veins – except for pulmonary veins – carry deoxygenated blood; arteries – except for pulmonary arteries – carry oxygenated blood.
3. *What is the systemic circulation?*

   The blood vessels that carry oxygenated blood from the left ventricle to body tissues and organs other than the lungs and return deoxygenated blood via the veins to the right atrium.
4. *What is the difference between muscular and elastic arteries?*

   Elastic arteries are large vessels with a diameter of up to 2.5 cm. They carry large volumes of blood away from the heart. The tunica media of elastic arteries contains a high density of elastic fibres and relatively few smooth muscle cells; as a result, elastic arteries are extremely resilient and can tolerate the pressure changes of the cardiac cycle. Muscular arteries are medium-sized arteries (diam 0.4 cm) (e.g. the axillary, brachial, radial, femoral, popliteal and tibial arteries). They distribute blood to the body's skeletal muscles and internal organs. They have thick tunica media layers, which contain more smooth muscle than elastic, enabling greater autonomic control of vasoconstriction and vasodilation.

5. *What is the function of the smooth muscle in blood vessel walls?*

   Smooth muscle allows for vasodilation (when the degree of muscle tone is decreased, e.g. smooth muscle at the entrance to capillaries regulate the amount of blood flow into each vessel. If the tissue becomes starved for oxygen, the smooth muscle cells relax, whereupon capillary diameter increases and so blood flow increases, delivering additional oxygen) and vasoconstriction. This mechanism allows the volume flow rate of blood to be altered.

6. *Name the three types of capillaries. What is the difference between them?*

   (a) Continuous capillaries: the endothelium forms a complete lining. These capillaries allow the diffusion of water, small solutes and lipid-soluble materials into the interstitial fluid; they prevent the loss of blood cells and plasma proteins.
   (b) Fenestrated capillaries: the endothelium has pores (windows) to allow the rapid movement of water and solutes (including peptides) between the plasma and interstitial fluid.
   (c) Sinusoids (sinusoidal capillaries): these resemble fenestrated capillaries that are flattened and irregularly shaped. In addition to being fenestrated, sinusoids commonly have gaps between adjacent endothelial cells. They allow the free exchange of water and solutes as large as plasma proteins and phagocytic cells (i.e. are "leaky").

7. *What would result if "Starling's law of the capillaries" does not hold?*

   If the same amount of fluid did not return to the capillaries at the venous end as the left at the arterial end (less than which returns via the lymphatic vessels), then the tissues would swell with retained fluid (become oedematous).

8. *(a) What causes blood to flow in the arterial side of the systemic circulation?*

   The pumping action of the heart.

   (b) What causes venous return (blood flow) in the venous side of the circuit?

   1. The presence of valves in the veins to prevent backflow ensures that blood keeps moving back to the heart.
   2. Venous return is assisted by skeletal muscle contraction, the squashing action of the skeletal muscle acting on veins pressing against the blood in the veins, forcing the valves open and driving blood towards the heart.

3. In the abdomen, the action of breathing creates a lower pressure in the thorax than in the abdomen, and this encourages blood to flow along the vena cava (blood from the head flows to the heart under gravity usually).

Exercise 14: Pressure

1. *Define pressure and explain how it is different from force.*

   Pressure is the amount of force (expressed in newtons) being exerted divided by the amount of area (expressed in square metres) on which the force is acting. The units of pressure are Pascals (Pa), which are equal to newtons per square metre. Since pressure is force divided by area, they are different (pressure is not a force). Furthermore, the same force may exert different pressures – a large pressure if the area is small or a small pressure if the area is large.

2. *Convert a blood pressure measurement of 130 mm Hg/80 mm Hg to units of kPa.*

   1 mm Hg = 0.133 kPa, so to convert mm Hg to kPa, the number of mm Hg is multiplied by 0.133: 130 mm Hg × 0.133 = 17.3, and 80 mm Hg × 0.133 = 10.6. So 130/80 (in mm Hg) is equivalent to 17.3/10.6 (in kPa).

3. *Use $P = F/A$ to determine the pressure exerted during CPR for the two cases below. In each case, the resuscitator can apply a force of 200 N.*

   (a) *If the whole hand (area = 140 cm^2 = 0.014 m^2) is used.*

$$200\,\text{N} \div 0.014\,\text{m}^2 = 14285.7\,\text{Pa} = 14.2857\,\text{kPa}.$$

   (b) *If only the "heel" of the hand is used (area = 40 cm^2 = 0.004 m^2).*

$$200\,\text{N} \div 0.004\,\text{m}^2 = 50000\,\text{Pa} = 50\,\text{kPa}.$$

Thus, a much greater pressure may be exerted (using the same force) if the heel of the hand is used rather than the palm!

4. *What is meant by positive pressure?*

   A pressure that is greater than atmospheric pressure (which is about 101 kPa); for example, a positive pressure of 10 kPa is in fact 10 kPa + 101 kPa.

5. *Write out Pascal's principle. Apply it to bed sore prevention or glaucoma or Queckenstedt's test or the knee or to a foetus surrounded by amniotic fluid.*

   Pascal's principle: the pressure exerted on an enclosed fluid is transmitted undiminished to all parts of the fluid and to its container's walls.
   PP applied to BSP: a patient lying on a water-filled mattress is exerting pressure on the enclosed water due to their weight. If the mattress cover is stretchy enough to follow the contours of the patient's body, then the mattress is in contact with the patient's entire lower surface. That is, the area over which the

patient's weight is acting is the entire lower surface of the patient rather than just their bony prominences (such as shoulder blades, buttocks, heels etc). This large area of contact means that the pressure exerted by the patient's weight is smaller and the chance of producing pressure sores is reduced.

PP applied to glaucoma: glaucoma occurs when the intraocular pressure is high. The eyeball is a bag of fluid. The pressure in the "bag" is transmitted to the walls of the bag. The blood vessels of the retina are in the wall. They are compressed by this pressure, and consequently, the cells of the retina are deprived of their blood supply. Damage to sight is the result.

PP applied to QT: Queckenstedt's test involves compressing the jugular vein, thus causing the venous sinus in the cranium to swell with blood (as the blood is prevented from draining away). The excess blood exerts an increased pressure in the cranium, which is passed onto the cerebrospinal fluid.

PP applied to the knee joint: the pressure in the knee joint can be distributed because of the enclosed synovial fluid.

PP applied to a foetus: a foetus is surrounded by amniotic fluid; hence, pressure on the abdomen of a pregnant woman is transmitted to the entire surface of the foetus and to the mother's abdominal wall via the amniotic fluid. Thus, the baby can be protected from high pressure.

6. *Use Henry's law to describe why breathing air with 30% O_2 is often a beneficial therapy.*

   Atmospheric air contains about 20% O_2, so 30% O_2 is air enriched with oxygen. Henry's law says that the amount of gas that dissolves in water is proportional to the pressure of the gas in contact with the water. (The higher is the partial pressure of gas adjacent to a liquid surface (e.g. in the lungs), the greater the amount of gas that will dissolve in the liquid). If the gas contains 30% O_2 (rather than 20% O_2), MORE oxygen will dissolve in the water; hence, breathing enriched air will allow more oxygen to dissolve in the alveolar fluid, so more oxygen will dissolve in the blood, and so more oxygen will be transported. This is beneficial for someone who has a gas exchange problem, is anaemic and has a pulmonary obstruction.

7. *Describe inhalation and exhalation using Boyle's law and pressure gradient.*

   Inhalation is achieved by contracting the diaphragm and/or the intercostal muscles. This causes the volume of the chest to increase. By Boyle's law, the pressure of the gas in the lungs will decrease to below that of the atmosphere. That is, there will be a higher air pressure outside the lungs than inside the lungs. This constitutes a pressure gradient. Consequently, the outside air will push its way into the lungs. On exhalation, the volume of the chest will decrease (as the diaphragm relaxes and the ribs recoil). By Boyle's law, a decreased volume will produce a higher pressure – higher than the pressure of the air in the atmosphere. This difference in pressure once again causes a flow of gas. Hence, "stale" air is pushed out of the lungs in the direction of the pressure gradient.

8. *In about four sentences, describe and explain the movement of fluid into and out of capillaries.*

The capillary hydrostatic pressure ranges from 35 mm Hg at the arterial end of a capillary to 18 mm Hg at the venous end. At the arterial end, blood enters the capillary from the arteriole at BP (capillary hydrostatic pressure) ~35 mm Hg, and this pressure is higher than blood colloid osmotic pressure ~25 mm Hg (which attracts water into the blood by osmosis); therefore, water flows out of the capillary at the arterial end. At the venous end, blood leaves via a venule at BP (capillary hydrostatic pressure) ~18 mm Hg, which is lower than blood colloid osmotic pressure ~25 mm Hg; hence, water flows into the capillary at the venule end.

Exercise 15: Blood Pressure and Control of Blood Pressure

1. *Define mean arterial blood pressure.*

$$\text{MAP} = \text{diastolic pressure} + 1/3\,\text{pulse pressure.}$$

The average pressure lies somewhere between the maximum (and momentary) systolic pressure and the minimum diastolic pressure.

2. *What are the three basic factors that determine arterial blood pressure?*

   (a) Cardiac output: the amount of blood ejected each minute by the left ventricle into the aorta; an increase in CO will increase the blood pressure.
   (b) Peripheral resistance: the resistance to blood flow due to friction between the blood and the vessel walls; it depends on the viscosity of the blood and the diameter and length of the vessels, especially the arterioles. The smaller the diameter of the vessel is, the more resistance it offers to blood flow. The greater the length of the blood vessels, the greater the resistance to flow.
   (c) Blood volume: decreased fluid volume within a closed system decreases the pressure on the vessels within that system.

3. *State the formula that relates peripheral resistance, stroke volume and heart rate to mean arterial blood pressure.*

$$\text{MAP} = \text{SV} \times \text{HR} \times \text{PR}$$

4. *What effect does vasoconstriction have on arterial blood pressure and why?*

Vasoconstriction raises blood pressure. The reason is that vasoconstriction or vasodilation can change the vessel diameter and, hence, change resistance to blood flow, therefore changing blood pressure. The smaller the diameter of blood vessels, the higher the resistance for the blood to pass; that is, the heart has to exert greater pressure.

5. *What are the effects on blood volume and arterial pressure of:*

   (a) *Increased glomerular filtration rate (GFR)?*

   Filtrate production is increased. Blood volume would decrease, and so would arterial pressure.

   (b) *Decreased GFR?*

   BV would increase, and BP would increase.

   (c) *An increase in sympathetic stimulation to the heart and vascular smooth muscle?*

   HR and the force of contraction would increase. Vasoconstriction would increase, so BP would increase.

   (d) *An increase in parasympathetic stimulation to the heart?*

   HR would decrease, the force of contraction would decrease, and so BP would decrease.

   (e) *ADH secretion?*

   ADH causes the walls of the collecting ducts of the kidney to become permeable to water. Consequently, water is re-absorbed and moves from the filtrate to the interstitial fluid. This ultimately causes blood volume to rise and hence blood pressure to rise.

   (f) *Drinking water?*

   As water is absorbed from the gastrointestinal tract, blood volume increases, and so does pressure.

6. *(a) What is the stimulus for renin release?*

   A fall in blood pressure.

   (b) How does angiotensin II arise from the release of renin?

   Renin converts angiotensinogen to angiotensin I; then in the capillaries in the lungs, an angiotensin-converting enzyme changes angiotensin I to angiotensin II.

   (c) What effects do angiotensin II have?

   (1) It causes adrenal glands (cortex) to release **aldosterone,** which increases the re-absorption of Na^+ from the filtrate by kidney tubules. (2) It causes the secretion of ADH by the posterior pituitary (hence, kidney tubules become permeable to water, which osmotically follows Na^+ out of the filtrate; therefore, urine production decreases as water is reclaimed from the filtrate and returns to blood – this maintains the circulating blood volume). (3) Angiotensin II stimulates thirst; we drink, which increases extracellular volume and hence

blood volume. (4) Angiotensin II stimulates cardiac output and is also a vasoconstrictor, in turn increasing systematic blood pressure.

(d) How do the drugs known as ACE inhibitors work?

They prevent the enzyme from acting to convert angiotensin I to angiotensin II; hence, blood pressure does not increase.

(e) Why do "calcium channel blockers" help reduce blood pressure?

It is because Ca $^{++}$ ions are necessary for muscle contraction, and these blockers block the entry of Ca^{++} ions to the vascular smooth muscle; hence, it cannot contract. Therefore, the arterioles dilate, causing the peripheral resistance to decrease, i.e. lower blood pressure.

7. *Name the major mechanisms for the short-term regulation of blood pressure.*

   The aortic reflex (a baroreceptor reflex) maintains systemic BP – baroreceptors located in the aortic arch. The carotid sinus reflex (a baroreceptor reflex) maintains adequate and constant blood flow in the brain – baroreceptors located in the carotid sinus.

8. *Write a paragraph about the role of the kidneys in the long-term control of blood pressure.*

   When arterial pressure falls, the juxtaglomerular cells of the kidney release renin, which catalyses the formation of angiotensin I, which is converted to angiotensin II by ACE. The effects of angiotensin II are the release of aldosterone, which increases the re-absorption of Na+ from the filtrate by kidney tubules and the secretion of ADH by the posterior pituitary; hence, kidney tubules are more permeable to water, which osmotically follows Na+ out of the filtrate. Angiotensin II is also a vasoconstrictor and stimulates thirst. The atrial natriuretic peptide is produced by muscle cells in the right atrium in response to the excessive stretching of atria; it reduces blood volume and blood pressure. The effect of ANP is it increases Na+ excretion in the kidneys and increases GFR, i.e. urine volume, so blood volume decreases and BP decreases. It also reduces thirst and blocks the release of ADH and aldosterone.

Exercise 16: Respiratory System

1. *Define the following: dead space, lung compliance and respiratory distress syndrome.*

   Dead space: that volume of air that is inhaled but does not reach the alveoli (air in the conducting zone of the bronchial tree) and hence does not participate in gas exchange.
   Lung compliance: the ease with which the lung can be expanded – low compliance means that the lung is difficult to expand and the greater is the force

required to expand the lungs. The greater the compliance, the easier it is to fill the lungs.

Respiratory distress syndrome: afflicts premature babies whose alveolar epithelial cells are not mature enough to produce the surfactant. It is required so that the surface tension of alveolar fluid is decreased to allow for easy expansion of the lung. Consequently, expanding their lungs without assistance is beyond their capability.

2. *How does the construction of the walls of the bronchi and bronchioles differ?*

   Bronchi have cartilage in their walls to hold them open. However, the walls of the primary, secondary and tertiary bronchi contain progressively less cartilage. Bronchioles have no cartilage in their walls – they do have smooth muscle so that they can constrict. The amount of tension in those smooth muscles has a great effect on bronchiole diameter and resistance to airflow.

3. *What can cause bronchodilation, and what can cause bronchoconstriction?*

   Bronchodilation happens when adrenalin or noradrenalin is present in the blood, CO_2 concentration goes up or the sympathetic nerves carry impulses to the lungs. Bronchoconstriction is triggered by parasympathetic nerve stimulation, by histamines released during allergic reactions and by irritants carried in air (smoke, dust and dander).

4. *Why does hyperventilation (= rapid shallow breaths) result in an increase in dissolved CO_2 in the blood?*

   Rapid shallow breathing will not ventilate the lungs adequately as a large proportion of each intake of breath will simply move into the dead space – shallow breaths do not have sufficient volume to refresh the alveolar volume with fresh air. Hence, CO_2 in the blood will not be "blown off". CO_2 then will accumulate in the blood and decrease blood pH.

5. *To what do the chemoreceptors in the respiratory centre of CNS respond? (And explain your answer.)*

   The respiratory centre responds to H^+. However, this ion cannot cross the blood-brain barrier. CO_2 from the blood can; when it enters the CSF, it dissolves to form carbonic acid, which in turn dissociates into bicarbonate and H+. The more CO_2 in the blood, the more H^+ will be formed in the CSF. So it might be said that CO_2 affects the chemoreceptors via H^+.

6. *How and why is the composition of alveolar air different from atmospheric air?*

   Alveolar air has less O_2 (because it is continually moving into the blood) and has more CO_2 (because it is continually moving out of the blood) and is saturated with water vapour (because it is in contact with moist membranes).

7. *What is FEV_1, and why is it decreased in obstructive diseases such as asthma?*

   It is the maximum volume of air that can be forcefully exhaled in one second (two or three attempts are allowed). It depends on the airways being open – i.e. unobstructed and unconstricted. The constriction caused by asthma means that the diameters of the airways are smaller; consequently, less air can pass through them. Healthy lungs can expel ≥80% of their volume in 1 s.

8. *Describe the chemical changes that occur in the RBC that facilitate carbon dioxide transport.*

   Of the CO_2 formed by cellular respiration and dissolved in plasma, 70% moves into the RBC where it reacts with water to form carbonic acid. This reaction is EXTREMELY rapid because it is catalysed by the enzyme carbonic anhydrase. The acid then dissociates into bicarbonate ions and hydrogen ions. The bicarbonate then diffuses out of the RBC and is transported back to the lungs in plasma (chloride ions swap places with bicarbonate ions so that the electrical potential stays the same on each side of the membrane; this ion exchange process does not require ATP). Hydrogen ions combine with oxyhaemoglobin, forming HbH+ and releasing oxygen in the process. Consequently, oxygen is unloaded in the tissues that are producing CO_2 – and this is good because that is just where oxygen is needed.
9. *State Henry's law. Use Henry's law to describe why breathing air with 30% O_2 is often a beneficial therapy.*

   Henry's law says that at a given temperature, the amount of gas that dissolves in water is proportional to the partial pressure of the gas in contact with the water.
   Atmospheric air contains about 20% O_2, so 30% O_2 is air enriched with oxygen. Henry's law says that the amount of gas that dissolves in water is proportional to the pressure of the gas in contact with the water. (The higher is the partial pressure of gas in contact with a liquid surface (e.g. in lungs), the greater will be the concentration of the dissolved gas in that liquid.) If the gas contains 30% O_2 (rather than 20% O_2), MORE oxygen will dissolve in the water. Hence, breathing enriched air, will allow more oxygen to dissolve in the alveolar fluid, so more oxygen will dissolve in the blood and more oxygen will be transported. This is beneficial for someone who has a gas exchange problem, is anaemic or has a pulmonary obstruction as gas exchange is enhanced. For anaemic patients (whose RBC concentration is lower), the extra amount of oxygen dissolved in plasma means more oxygen is available to the tissues than would be the case if they were breathing 20% oxygen.

10. *State Boyle's law. Describe inhalation and exhalation using Boyle's law and pressure gradient.*
    See Homework Exercise 14, question 7.

For gas in a closed container and at a constant temperature, pressure (P) is inversely proportional to volume (V); this relationship, which can be presented as $P = 1/V$, is Boyle's law.
Inhalation is achieved by contracting the diaphragm and/or the intercostal muscles. This causes the volume of the chest to increase. Based on Boyle's law, the pressure of the gas in the lungs will decrease to below that of the atmosphere. That is, there will be higher air pressure outside the lungs than inside the lungs. This constitutes a pressure gradient. Consequently, outside air will push its way into the lungs. On exhalation, the volume of the chest will decrease (as the diaphragm relaxes and the ribs recoil). As per Boyle's law, a decreased volume will produce a higher pressure – higher than the pressure of air in the atmosphere. This difference in pressure once again causes a flow of gas. Hence, "stale" air is pushed out of the lungs in the direction of the pressure gradient.

11. *What are the functions of the mucous glands and the ciliated epithelial cells?*

    Mucous glands secrete mucous onto the surface of large airways. Mucous traps particles suspended in inhaled air (also contains immunoglobulins to disable pathogens). Ciliated epithelial cells have cilia that wave the mucous towards the top of the bronchial tree so that it may be swallowed (then stomach acid acts on any pathogens).

12. *Give definitions for the following: eupnoea, apnoea, bradypnoea, tachypnoea and dyspnoea.*

    Eupnoea – normal breathing (12–20 bpm).
    Apnoea – cessation of breathing (may be temporary).
    Bradypnoea – abnormally slow breathing (? <10 bpm).
    Tachypnoea – rapid breathing (? >20 bpm) may be due to exercise.
    Dyspnoea = subjective sensation of difficulty in breathing and response to that sensation (= shortness of breath, breathlessness).

13. *Why would you encourage an anxious person to breathe more deeply?*

    Deep breaths increase the volume of air intake – this decreases the ratio of "dead air" (expired air from the alveoli that is still in the airways after exhalation and is then drawn back into the lungs with a new breath) to fresh air taken into the alveoli. Thus, a deep breath draws more fresh air (with more oxygen and less carbon dioxide) into the alveoli, and this improves gas exchange across the respiratory membrane (which in turn will calm the subject as they decrease CO_2 in their blood).

14. *What would you anticipate would be the effect of pneumothorax (air in the intrapleural space) on breathing?*

    A pneumothorax takes up space within the chest cavity, so lungs cannot expand as fully as they should. This leads to decreased gas exchange and "difficulty in breathing" and perhaps increased CO_2 in the blood and respiratory acidosis. If the lung collapses, breathing difficulty will increase.

15. *What could you anticipate would be the effect on breathing of a spinal cord injury at the level of the sixth cervical vertebra?*

    If spinal cord is severed at C6, then the spinal nerves that leave the spinal cord below C6 will not carry impulses (this results in quadriplegia). So intercostal muscles of the chest will not work in breathing (because the spinal nerves will not function). However, the diaphragm is innervated by the phrenic nerve, which leaves the SC (via cervical plexus) from C3, C4 and C5. Hence, the diaphragm will still work. The subject will be able to breathe unassisted using the diaphragm and so will not die of suffocation at the trauma site. (Severing the SC above C3 means lung ventilation (breathing) will cease.)

Exercise 17: Reproductive System

1. *List the structures of the male reproductive system in the order that a spermatozoon would pass through them.*

   Testis (seminiferous tubule – straight tubules, rete testis – efferent ductules), epididymis, ductus (vas) deferens, ejaculatory duct, urethra (and out through the penis via the external urethral meatus).
2. *What is the function of testosterone in males?*

   (a) It stimulates spermiogenesis (in the sustentacular cells of the testes).
   (b) It Affects CNS function (libido and male behaviour).
   (c) It stimulates metabolism (masculine body shape, muscle growth).
   (d) It establishes and maintains secondary sex characteristics (hair, muscle mass, body size, fat deposits, thicker vocal cords, Adam's apple, larger hands and feet, thicker skin).
   (e) It maintains glands and organs of the reproductive tract.

3. *Where do spermatozoa physically mature, and where do they become "capacitated"?*

   They physically mature in the epididymis but are still immobile.
   They become capacitated in (1) the ejaculatory duct (after mixing with the secretions of the seminiferous glands and (2) when exposed to conditions in the female reproductive tract.

4. *What happens in the ejaculatory duct?*

   Spermatozoa capacitation: secretions of the seminal glands are added to sperm and fluid from the epididymis; sperm become motile.

5. *What is the composition of semen?*

   Typical ejaculate is 2–5 ml. Semen typically contains spermatozoa (20 million to 100 million spermatozoa per millilitre of semen), seminal fluid (a mixture of glandular secretions with a distinct ionic and nutrient composition) and enzymes. It also contains electrolytes, protein, fructose, lipids, vitamin C, water and mucus (caritene, prostaglandins)

6. *Write a paragraph that summarises the events of spermatogenesis. (Do not merely copy out the textbook!)*

   The complete process of spermatogenesis takes about 64 days. It involves three processes:

   1. Mitosis: stem cells undergo cell divisions. One daughter cell from each division remains in place, while the other is pushed towards the lumen of the seminiferous tubule, and the displaced cells differentiate into primary spermatocytes, which prepare to begin meiosis.
   2. Meiosis: it is a special form of cell division involved in gamete production. Human gametes contain 23 chromosomes, half the amount found in somatic cells. As a result, the fusion of the nuclei of a male gamete and a female gamete produces a cell that has a normal number of chromosomes.
   3. Spermiogenesis: spermatids are small unspecialised cells. In spermiogenesis, spermatids differentiate into physically mature spermatozoa, which are among the most highly specialised cells in the body.

7. *What is the function of estradiol?*

   (a) Stimulates bone and muscle growth (female pattern)
   (b) Maintains female secondary sex characteristics (hair distribution, breasts, location of adipose tissue, voice, pelvis)
   (c) Affects the CNS (sexual drive and female behaviour)
   (d) Maintains accessory reproductive glands/organs
   (e) Initiates growth and the repair of the endometrium
   (f) Promotes the development of breasts

8. *Summarise the hormonal control of the ovarian cycle up to the luteal phase.*

   (a) Release of GnRH: the cycle begins with the release of GnRH by the hypothalamus (at 1 pulse every 60–90 min), which stimulates the production and secretion of FSH (from the anterior lobe of the pituitary gland) and the production of LH.
   (b) Follicular phase of the ovarian cycle: FSH acts on the ovary to stimulate follicle development, and the developing follicles produce oestrogen (which inhibits LH release) and secrete inhibin, which causes the FSH level to decrease due to negative feedback effects.
   (c) Luteal phase: as one or more tertiary follicles begin to form, the concentration of circulating oestrogen rises steeply; hence, GnRH pulse frequency increases to 36 per day. Around day 10 of the cycle, the combination of increased GnRH pulse frequency and elevated oestrogen levels stimulates LH secretion (the effect of oestrogen on LH secretion changes from inhibi-

tion to stimulation). On around day 14, a massive LH level triggers the completion of meiosis I, ovulation and the formation of the corpus luteum. The corpus luteum secretes progesterone, which stimulates and sustains endometrial development. After ovulation, progesterone levels rise and oestrogen levels fall, which suppress GnRH secretion. If pregnancy does not occur, the corpus luteum will degenerate after 12 days, and as the progesterone level decreases, GnRH secretion increases, and a new cycle begins.

9. *What are the male secondary sex characteristics?*

   Hair distribution (beard), thicker skin, thicker vocal cords and deeper voice (than females), body shape (larger hands and feet, broad shoulders, narrow hips, muscle development), body size, fat distribution (around the abdomen) and Adam's apple.

10. *What are the female primary sex characteristics?*

    The ovaries, the reproductive tract (uterine tubes (fallopian tubes), uterus, vagina) and external genitalia.

11. *Define menstrual cycle, menarche and menopause.*

    The menstrual cycle is a repeating series of changes in the structure of the endometrium (~28–day cycle); it is the preparation of the uterus (proliferation of the uterine epithelium) to receive the fertilised ovum and the subsequent destruction of tissue (and the passing of blood and debris out through the vagina) if no embryo implants.
    Menarche is the onset of a female's first menstrual cycle (the first cycle).
    Menopause is the ultimate cessation of the menstrual cycle as a result of a woman's ageing.

12. *Outline the major changes in GnRH, FSH, LH, estradiol and progesterone before and after ovulation.*

    Similar to Q8.
    Before ovulation: GnRH stimulates the production and secretion of FSH and the production of LH. FSH stimulates follicle development, and developing follicles produce oestrogen, which inhibits LH, and inhibin, which causes FSH levels to decline. An increase in oestrogen causes the pulse frequency of GnRH release to increase to 1 pulse/30–60 mins. Persistently high oestrogen combined with a high level of GnRH also stimulates LH secretion.
    After ovulation: the corpus luteum releases progesterone – which prepares the uterus for pregnancy (and some oestrogen). As the progesterone level rises and oestrogen falls, GnRH pulse frequency declines to 1–4 pulses/day. This stimulates LH secretion (which maintains CL) more than it does FSH. The progesterone levels remain high for the next week, but unless pregnancy occurs, the CL begins to degenerate. As progesterone and oestrogen levels fall markedly, the GnRH pulse frequency increases, and the cycle begins again.

13. *What is the function of GnRH and the result of its release in males and females?*

GnRH (produced in the hypothalamus) stimulates the release of FSH and LH from the anterior pituitary (in both males and females).
In males: FSH targets sustentacular (Sertoli) cells (in the testes), which (in the presence of testosterone) promote spermiogenesis. LH targets interstitial (Leydig) cells (of the testes), which produce testosterone.
In females (at 1 pulse/60–90 min): GnRH causes FSH release from the anterior pituitary. FSH targets the ovary and stimulates follicle development. At the pulse/30–60 min, GnRH, together with oestrogen, stimulates LH release from the anterior pituitary. This "surge" in LH stimulates the completion of meiosis I by oocytes and the ovulation and development of the corpus luteum.